T0138941

Carbon-Based Electronics

Carbon-Based Electronics

Transistors and Interconnects at the Nanoscale

Ashok Srivastava
Jose Mauricio Marulanda
Yao Xu
Ashwani K. Sharma

PAN STANFORD PUBLISHING

Published by

Pan Stanford Publishing Pte. Ltd.
Penthouse Level, Suntec Tower 3
8 Temasek Boulevard
Singapore 038988

Email: editorial@panstanford.com
Web: www.panstanford.com

British Library Cataloguing-in-Publication Data
A catalogue record for this book is available from the British Library.

**Carbon-Based Electronics: Transistors and Interconnects
at the Nanoscale**

ISBN 978-981-4613-10-1 (Hardcover)
ISBN 978-981-4613-11-8 (eBook)

Printed in the USA

Contents

Preface

The past 50 years have witnessed notable advancements in the field of very large scale integrated circuits with shrinking transistor geometries as predicted by Moore's law and applications in all walks of our lives. We are still guided by Moore's law. However, we are reaching the end of the curve close to year 2020 imposed by the laws of physics. Semiconductor Research Corporation in its past 2003 International Technology Roadmap ofSemiconductors report has referred to several non-classical devices, including those based on carbon nanotubes, which could be the candidates of future technology as the end of Moore's law approaches. The interconnects in sub-nanometer CMOS technology nodes are already facing problems in copper interconnect due to increase in its resistance.

Since the discovery of carbon nanotubes in 1991 by Japanese physicist Dr. Sumio Iijima, voluminous research has been done in the field of one-dimensional carbon nanotube material for numerous applications, including those for the possible replacement of silicon used in the fabrication of CMOS chips. One of the interesting features of carbon nanotube is that it can be metallic or semiconducting with a bandgap depending on its diameter. Since carbon nanotubes are planar graphene sheets wrapped into tubes, electrical characteristics vary with the tube diameter and the wrapping angle of the graphene. Carbon nanotubes can be manipulated in a controlled way in their position, shape, and orientation with the use of the atomic force microscope. Carbon nanotube as an interconnect material and its integration with CMOS process offers the much-awaited solution. In search of non-classical devices and related technologies, both the carbon nanotube–based field-effect transistors and metallic carbon nanotube interconnects are being explored extensively for emerging logic devices for very large scale integration.

Transistors in the integrated circuit design for analog, digital, mixed-signal applications, including those for radio frequency operation, require equivalent circuit models for use with device- and circuit-level simulators. Although various models for carbon nanotube–based transistors and interconnects have been proposed in the literature, an integrated approach and compatibility with present simulators are yet to be made available. This book present the material that is an attempt in this direction, where models for both transistors and interconnects based on carbon nanotubes are developed from the fundamental understanding of the material and solid-state physics and made compatible with commercial integrated circuit design simulators through Verilog-AMS codes for design and analysis of integrated circuits. A need of such an approach motivated the authors to begin developing a better understanding of current transport in metallic and semiconducting carbon nanotubes and building closed-form analytical models for the design and analysis of carbon nanotube–based integrated circuits.

An overview of single-walled, multi-walled, and bundle of single-walled carbon nanotubes, electrical properties, and carbon nanotube–based transistors, interconnection, and integrated circuit is presented in Chapter 1. In Chapter 2, current transport phenomenon in semiconducting carbon nanotubes is studied from the understanding of physics of semiconductors and closed-form analytical equations are derived. Chapter 2 serves as a basis of developing current transport model of carbon nanotube field-effect transistors in Chapter 3 similar to MOSFET models for the analysis and design of integrated circuits. The developed analytical current transport models verifying established experimental current–voltage characteristics have been used in design of basic logic gates. Integrated circuits based on carbon nanotube field-effect transistors will also need to integrate carbon nanotube metallic interconnect. In Chapter 4, a detailed study has been conducted on current transport in metallic single-walled carbon nanotubes and one-dimensional fluid model has been developed, which accounts for the electron–electron interaction, and compared with the two-dimensional fluid model. The fluid model for single-walled carbon nanotube as interconnect has been extended in Chapter 5 for multi-walled and bundle of single-walled carbon nanotubes as interconnects. Suitability of carbon nanotube, single-walled, multi-walled, and bundle of single-walled as interconnect for ballistic transport and local and global interconnection is investigated and compared with the performance with copper as interconnects, and S-parameters are studied in detail, including the power dissipation.

Phase-locked loops are widely used in high-speed and high-frequency data communication systems. Phase noise is one of the major causes of concerns and originates mainly from one of its building blocks, the voltage-controlled oscillator, which requires high-Q inductors. Carbon nanotube wire has reduced skin effect compared to metal conductors such as the copper and has a great promise for the realization of high-Q on-chip conductors. In Chapter 6, a model of carbon nanotube as an inductor is studied, and a feasible design of a very high frequency LC voltage-controlled oscillator is presented using multi-walled and single-walled bundle wires as an inductor in the LC tank circuit. The observed reduced phase noise in voltage-controlled oscillator makes the high-Q carbon nanotube wire inductors a very promising component for RF circuit design. Energy recovery techniques are playing an important role in modern electronic circuit design due to urgent need of low power dissipation in portable communication and computing systems. Although energy recovery techniques for CMOS circuits are well developed, such techniques are not fully explored for carbon nanotube field-effect transistor–based circuits. In Chapter 7, we have explored design and analysis of basic energy recovery carbon nanotube field-effect transistor circuits based on models developed in Chapter 3 and CMOS-based energy recovery design styles. We have shown that much work is needed to reduce the on-chip power density above 1 GHz. While developing current transport models for transistors and interconnects based on carbon nanotubes for integrated circuit design, it was observed that the compatibility of models with simulator such as SPICE is too difficult and time consuming due to the complexity of models. Instead, Verilog Analog Mixed-Signal (Verilog/AMS) solves the much complex problem and fewer steps are required for simulations. This is because model equations for new devices can be put into the Verilog-AMS coding and simulator will call the code. Chapter 8, the last chapter, concludes with Verilog/AMS codes for the carbon nanotube field-effect transistor in Cadence and usefulness demonstrated.

This book can serve as a textbook for graduate-level course in nanoelectronics for students in electrical and electronics engineering, applied physics, and materials science with focus on reduced-dimension materials such as carbon nanotubes. The material covered in the book will be very useful for graduate students for thesis and dissertation research exploring the integration of carbon nanotubes as interconnects with sub-nanometer CMOS technology nodes, high-frequency and high-speed electronics, carbon nanotube field-effect transistor sensor electronics, and numerous other applications. The book will be very useful for practicing engineers in the field of VLSI design and technology, semiconductors, nanoscience and nanotechnology, and microelectronics.

The authors would like to thank Drs. Yang Liu and Rajiv Soundararajan for the design and testing of carbon nanotube–based integrated circuits in Cadence using developed models for field-effect transistors and wire inductors. The authors gratefully acknowledge Mr. Clay Mayberry of the United States Air Force Research Laboratory, New Mexico for his strong support, encouragement, and useful discussions. The book would have not been complete without the encouragement and support of our families. We would like to give special thanks to our families: to my wife, Shashi, son, Siddharth, daughter, Gitanjali, son-in-law, Saurabh, grandson, baby Ayaan, and my old mother back in India for their continued support and encouragement (Srivastava); to my parents and sisters for their constant encouragement and support (Marulanda); to my wife and sons for their support (Xu); and my daughter, Anna, son, Andrei, and my father, Mohinder, for their support and encouragement during the course of the writing of the book (Sharma).

Ashok Srivastava
Jose Mauricio Marulanda
Yao Xu
Ashwani K. Sharma

Chapter 1

Introduction to Carbon Nanotubes

Semiconductor Research Corporation in its International Technology Roadmap of Semiconductors report (ITRS 2003) has referred to several non-classical devices, which could be the candidates of future technology to replace the existing silicon MOSFETs as the end of Moore's law approaches year 2020 [1]. Double-gate MOSFET and FinFET are recognized as two of the most promising candidates for future very large scale integrated (VLSI) circuits [2–5]. The carbon nanotube field-effect transistor (CNT-FET) is regarded as an important contending device to replace silicon transistors [6–7] since many of the problems that silicon technology is facing are not present in CNTs. For example, carrier transport is 1-D in carbon nanotubes; the strong covalent bonding gives the CNTs high mechanical and thermal stability and resistance to electromigration; and diameter is controlled by its chemistry and not by the standard conventional fabrication process [2].

For interconnects, as CMOS processes scale into the nanometer regime, lithography limitations, electromigration, and the increasing resistivity and delay of copper interconnects have driven the need to find alternative interconnect solutions [8]. Carbon nanotubes have emerged as a potential candidate to supplement copper interconnects because of their ballistic transport and ability to carry large current densities in the absence of electromigration [9]. Previous studies that assess the potential use of CNTs as

Carbon-Based Electronics: Transistors and Interconnects at the Nanoscale
Ashok Srivastava, Jose Mauricio Marulanda, Yao Xu, and Ashwani K. Sharma
Copyright © 2015 Pan Stanford Publishing Pte. Ltd.
ISBN 978-981-4613-10-1 (Hardcover), 978-981-4613-11-8 (eBook)
www.panstanford.com

interconnects [10–13] primarily focus on the relative interconnect delay of CNTs to copper for sub-nanometer CMOS technology nodes. Carbon nanotubes are being explored extensively as the material for making future complementary devices, integrated circuits [14–16], interconnects [17], and hybrid CMOS/nanoelectronic circuits [18]. In the following, some insight into the carbon nanotube as a material, realization of field-effect transistor and interconnection for integrated circuits will be presented.

1.1 Introduction—Carbon Nanotubes

In 1960, Bacon of Union Carbide [19] reported observing straight hollow tubes of carbon that appeared as graphene layers of carbon. In 1970s, Oberlin et al. [20] observed these tubes again by a catalysis-enhanced chemical vapor deposition (CVD) process. In 1985, random events led to the discovery of a new molecule made entirely of carbon, 60 carbons arranged in a soccer ball shape [21]. In fact, what had been discovered was an infinite number of molecules: the fullerenes, C_{60}, C_{70}, C_{84}, etc., every molecule with the characteristic of being a pure carbon cage. These molecules were mostly seen in a spherical shape. However, it was until 1991 that Iijima [22] of NEC observed a tubular shape in the form of coaxial tubes of graphitic sheets, ranging from two shells to approximately 50. Later, this structure was called multi-walled carbon nanotube (MWCNT). Two years later, Bethune et al. [23] and Iijima and Ichihashi [24] managed to observe the same tubular structure, but with only a single atomic layer of graphene, which became known as a single-walled carbon nanotube (SWCNT).

Most SWCNTs have a diameter close to 1 nm, with a tube length that can be many thousands of times longer. The structure of an SWCNT can be conceptualized by wrapping an atomic thick layer of graphite called graphene into a seamless cylinder. Multi-walled carbon nanotubes consist of multiple layers of graphite rolled in to form a tubular shape. Since CNTs are planar graphite sheets wrapped into tubes, electrical characteristics vary with the tube diameter and the wrapping angle of graphene [25]. One of the interesting features of the carbon nanotube is that it can be metallic or semiconducting with bandgap depending on its chirality [25–28].

There are four types of natural occurring carbon: diamond, graphite, ceraphite, and fullerenes. Fullerenes are molecules formed entirely of carbon and take the shape of a hollow sphere, ellipsoid, or a tube. Fullerenes that take the shape of a tube are called buckytubes or nanotubes. Carbon nanotubes can be pictured as a result of folding graphene layers into a tubular structure as seen in Fig. 1.1 [29]. These cylindrical forms of carbon nanotubes can be single-walled or multi-walled depending on the number of shells that form the tubular structure [28]. Single-walled carbon nanotubes are composed of one shell of carbon atoms. Multi-walled carbon nanotubes have multiple nested shells of carbon atoms, as shown in Fig. 1.2. Single-walled carbon nanotubes tend to adhere strongly to each other forming ropes or bundles of nanotubes as shown in Fig. 1.3 [29] exhibiting physical properties of both metallic and semiconducting materials [29].

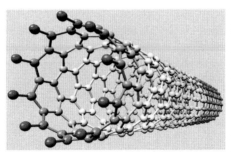

Figure 1.1   Single-walled carbon nanotube (Nanotube Modeler Software) [29].

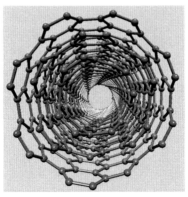

Figure 1.2   Multi-walled carbon nanotubes (Nanotube Modeler Software) [29].

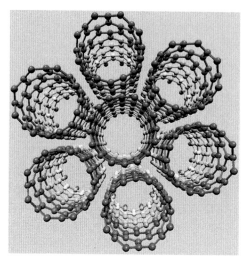

Figure 1.3 Single-walled carbon nanotubes bundle (Nanotube Modeler Software) [29].

Carbon nanotubes exhibit promising mechanical and electrical properties. Tables 1.1 and 1.2 summarize mechanical and electrical properties of carbon nanotubes [30–32]. They also compare the properties with that of silicon, currently used material in CMOS technologies.

Table 1.1 Mechanical properties of carbon nanotubes and a comparison with other materials [30]

Material	Young's modulus (TPa)	Tensile strength (GPa)	Elongation at break (%)	Thermal conductivity (W/mK)
SWCNT	1–5	13–53	16	3500–6600
MWCNT	0.27–0.95	11–150	8.04–10.46	3000
Stainless steel	0.186–0.214	0.38–1.55	15–50	16
Kevlar	0.06–0.18	3.6–3.8	~2	~1
Copper	0.11–0.128	0.22		385
Silicon	0.185	7		149

Table 1.2 Electrical properties of carbon nanotubes and comparison with other materials [31,32]

	Semiconductor				Metal		
	Semiconducting				Metallic		
Parameter	SWCNT	Silicon	GaAs	Ge	Parameter	SWCNT	Copper
Bandgap (eV)	0.9/diameter	1.12	1.424	0.66	Mean free path (nm)	1000	40
Electron Mobility (cm^2/Vs)	20000	1500	8500	3900	Current density (A/cm^2)	10^{10}	10^6
Electron Phonon Mean free path (Å)	~700	76	58	105	Resistivity ($\Omega \cdot m$)	~10^{-5}	1.68×10^{-8}

Metallic carbon nanotubes conduct extremely large amount of current densities. This property is what allows the application of metallic CNTs in interconnection substituting to metal wires, such as copper, for the next generation of integrated circuits. On the other hand, semiconducting CNTs can be switched on and off by using a gate electrode. This property is what allows the application of CNTs in implementing field-effect transistors.

Single-walled carbon nanotubes have risen as one of the most likely candidates for miniaturizing electronics beyond current technology. The most fundamental application of metallic SWCNTs is in interconnection. Since it is difficult to prepare metallic CNT, improved fabrication and process integration of metallic MWCNTs and CNT bundles interconnects have been reported [33,34].

1.2 CNT-Based Integrated Circuits

Studies have demonstrated that semiconducting CNTs have excellent electrical properties, including long mean free path (~0.7 μm) and high electron mobility [31,35,36]. Experiments with CNT-FETs [7,15,37] have further demonstrated that transistors based on semiconducting CNTs have large transconductance, which indicates a great potential for sub-nanometer integrated circuits as demonstrated through the fabrication of five-stage ring oscillator circuit by Chen et al. [38].

It has been suggested that high-κ dielectrics are essential for future transistors due to low leakage currents and reduced power dissipation [39,41]. However, a fundamental problem for conventional semiconductors is the degradation of electrical properties due to carrier scattering mechanisms introduced at the high-κ film semiconductor interface [42]. Recently, Javey et al. [43–46] have shown that CNT-FETs can be operated in ballistic range with high-κ dielectrics, thereby opening the door to ultrafast devices since both the ballistic transport of electrons and high-κ dielectrics facilitate high on-current that is directly proportional to the speed of a transistor.

Early CNT-FETs were fabricated on oxidized Si substrates [7,47]. The poor gate coupling due to the thick SiO_2 layer and back gate geometry limited their applications. However, in 2002, the implementation of top-gate geometry [15,16] made the CNT-FET a more promising candidate for next-generation field-effect transistors. Both n-type CNT-FETs and p-type CNT-FETs were made [46] and demonstrated for performance similar to current MOSFETs. Additional improvements in the metal–CNT contact resistance at the source and drain ends have led improved CNT-FET performances [45,48]. Now with improved processing techniques, CNT-FETs with very high ON/OFF switching ratio and high carrier mobility have been fabricated [49–53]. In addition, fabrication of inverters that are composed of n-type and p-type CNT-FETs were also reported in [54,55]. Recently, Zhang et al. [56] proposed a doping-free fabrication of CNT-based ballistic complementary metal-oxide semiconductor (CMOS) devices and circuits, which are compatible with current CMOS fabrication processes. This work may lead to the fabrication of complex CNT-based integrated circuits.

Metallic CNTs have attracted significant attention because their current carrying ability is remarkable. Ballistic transport of electrons has been observed and values for the conductance that approaches the theoretical limit $(4e^2/h)^1$ [31] have been measured at small biases [57]. Metallic CNTs hold promise as interconnects in sub-nanometer CMOS circuits because of their low resistance and strong mechanical properties. An emerging problem

---

[1]$4e^2/h = 155$ μS

with interconnects in sub-nanometer CMOS technology is the breakdown of copper wires due to electromigration when current densities exceed 10^6 A/cm^2 [58]. Preliminary work [59,60] have shown that an array of nanotubes can be integrated with silicon technology and holds promise as vertical vias to carry more than an order of magnitude larger current densities than conventional vias. Wei et al. [9] have demonstrated that MWCNTs can carry current densities approaching 10^{10} A/cm^2. Metallic CNTs are excellent wires, with near-perfect experimentally measured conductance. This is because surface scattering, disorder, defects and phonon scattering, which lead to a decrease in conductance, have negligible effects in metallic CNTs, especially when the size of the conductor is shrinking. The reasons for this are the following.

The acoustic phonon mean free paths in CNTs are longer than a micron [61]. The dominant scattering mechanisms are due to zone boundary and optical phonons with energies of approximately 160 and 200 meV; but scattering with these phonons at room temperature is ineffective at small biases [62]. In a silicon field-effect transistor, there is significant scattering of electrons due to the disordered nature of the Si–SiO$_2$ interface. However, the CNT has a crystalline surface without disordered boundaries [62].

Any potential that is long-ranged compared with the CNT lattice constant will not effectively couple the two crossing sub-bands because of lack of wave vector components in the reciprocal space [63]. The electrons in the crossing sub-bands of carbon nanotubes have a large velocity of 8×10^5 m/s at the Fermi energy. There are only two sub-bands at the Fermi energy. These two facts make the electron reflection probability due to disorder and defects small [64,63].

1.3   CNT-Based Circuit Modeling

As discussed earlier, CNT-FETs and metallic CNT wires show performance metrics significantly above those of Si MOSFETs and metal interconnect wires, respectively. Liang et al. [65] fabricated SWCNT FETs in which MWCNTs are used as local interconnects, making a successful first step towards CMOS circuits fabricated entirely from carbon nanotubes. However, studies of individual CNT-FET and CNT wire are not comprehensive enough to enable conclusions about development of large-scale CNT-based integrated circuits. The viability of CNT-FET circuits depends on the behavior of logic gates that are composed of multiple CNT-FETs and used in larger scale circuits. Therefore, circuit models including device models and interconnects model are necessary to predict the behavior of CNT-based circuits. Furthermore, like the great help by the CMOS models in optimization of CMOS circuits design, it will be helpful to optimize the CNT-based circuits design by utilizing the CNT-FET and CNT interconnect models. As a result, CNT-FET modeling and CNT interconnects modeling are the current focus of most active research in CNT-based integrated circuits design.

This work will focus on CNT-FETs and CNT interconnects made using semiconducting and metallic SWCNTs, respectively, and on some basic circuits, such as the inverter and ring oscillator, which are composed of complementary CNT-FETs and CNT interconnects.

1.3.1 CNT-FET Modeling

A good amount of work on modeling CNT-FETs has been reported [10,15,45,47,66–74]. However, these models are still numerical-based and require a mathematical/software realization. Recently, Srivastava et al. [75] have obtained an analytical solution of current transport model for the CNT-FET for analysis and design of CNT-FET-based integrated circuits.

(1) A simplified density of states function for carbon nanotubes has been experimentally demonstrated by Mintmire and White [76]. We have used these results in Chapter 2 to analytically derive the carrier concentration in carbon nanotubes. The intrinsic carrier concentration and doping effects are investigated for different carbon nanotubes. Results for varying diameters and wrapping angles are also presented.

(2) In Chapter 3, we utilize basic physical properties of field-effect transistors [77,78] and the work on carrier concentration of carbon nanotubes [79] to present analytical modeling equations describing the current transport in CNT-FETs. I–V characteristics for normal and subthreshold operation are demonstrated and their dependence on the chiral vectors and device geometries are investigated. In deriving the I–V characteristics of CNT-FETs, analytical model equations for threshold voltage (V_{th}) and saturation voltage ($V_{ds,sat}$)

are each derived in the process. These derivations are also investigated in much greater detail and the respective correlation to the chiral vectors and device geometry of CNT-FETs is demonstrated. In this chapter, we also use Verilog-AMS language to describe the developed non-ballistic CNT-FET models for simulations in Cadence/Spectre and study the CNT-based circuits, such as inverter pair, ring oscillator.

(3) We have also focused on the applications of our derived modeled equations. The results obtained from the *I–V* characteristics are used to generate voltage transfer characteristics of basic logic devices based on complementary CNT-FETs. We use our small signal equivalent circuit model [80–81] and current transport modeled equations [80–82] for CNT-FETs to study and analyze the frequency response of CNT-FETs and CNT-based logic devices and to establish a dependence on the chiral vectors and device geometry making predictions for very high frequency behavior in integrated circuit applications.

1.3.2 CNT Interconnect Modeling

A model describing the electromagnetic field propagation along a CNT is indispensable in order to study the interconnection performance of CNT while comparing with traditional metal interconnects. Three theories are used to build different models. Lüttinger liquid theory [83] describes interacting electrons (or other fermions) in one-dimensional conductor and is necessary since the commonly used Fermi liquid model breaks down in one-dimension. Burke [84,85] regards that electrons are strongly correlated when they transport along the CNT and proposed a transmission line model based on the Lüttinger liquid theory. Another transmission line model was built based on the Boltzmann transport equation (BTE) [86]. Two-dimensional electron gas, where the charged particles are confined to a plane and neutralized by an inert uniform rigid positive plane background was studied by Fetter [87,88]. Based on the work of Fetter [87,88], Maffucci et al. [89] investigated electron transport along the CNT and proposed a third model, fluid model. The first model is based on quantum dynamics concepts; the second model requires solving the Boltzmann transport equation; the third model has been developed within the framework of the classical electrodynamics and is simple on concepts and mathematical modeling.

The organization of the book is as follows. In Chapter 1, an overview of carbon-based electronics is presented in context of emerging devices and integrated circuits. In Chapter 2, current transport in semiconducting CNTs is modeled analytically, which forms the basis of Chapter 3, where current transport is developed for CNT-FETs. It is shown that the model can be integrated with EDA tools such as Cadence/Spectre for CNT-FET-based circuit design and simulation. In Chapter 4, we modify the two-dimensional fluid model of electron transport in carbon nanotubes to include electron–electron repulsive interaction and built a semi-classical one-dimensional fluid model. By using this model, we calculate the transmission line parameters for the metallic SWCNT. We also applied the one-dimensional fluid model to study the MWCNT and SWCNT bundle interconnect wires and analyzed their performances in Chapter 5. In Chapter 6, we have proposed the use of high-Q on-chip CNT wire inductors in design of LC-voltage controlled oscillators for phase-locked loop systems. In Chapter 7, we introduce some CMOS concepts in CNT-based circuits, such as the energy recovery techniques to study the performances of CNT based circuits. Chapter 8 concludes with the writing of Verilog-AMS codes in Cadence, one of the EDA tools for CNT-FET models in design and analysis of carbon nanotube based integrated circuits.

References

1. (2003). International Technology Roadmap for Semiconductors. Available: http://www.itrs.net/Links/2003ITRS/Home2003.htm.

2. Frank, D. J., Dennard, R. H., Nowak, E., Solomon, P. M., Taur, Y., and Hon-Sum Philip, W. (2001). Device scaling limits of Si MOSFETs and their application dependencies, *Proc. IEEE*, **89**, 259–288.

3. Wong, H. S. P., Chan, K. K., and Taur, Y. (1997). Self-aligned (top and bottom) double-gate MOSFET with a 25 nm thick silicon channel, *IEDM Tech. Dig.*, 427–430.

4. Huang, X., Lee, W.-C., Kuo, C., Hisamoto, D., Chang, L., Kedzierski, J., Anderson, E., Takeuchi, H., Choi, Y.-K., Asano, K., Subramanian, V., King,

T.-J., Bokor, J., and Hu, C. (1999). Sub 50 nm FinFET: PMOS, *IEDM Tech. Dig.*, 67–70.

5. Iwai, H. (2007). Future of silicon integrated circuit technology, *Proc. International Conference on Industrial and Information Systems*, pp. 571–576.

6. Avouris, P., Appenzeller, P., Martel, J. R., and Wind, S. J. (2003). Carbon nanotube electronics, *Proc. IEEE*, **91**, 1772–1784.

7. Tans, S. J., Verschueren, A. R. M., and Dekker, C. (1998). Room-temperature transistor based on a single carbon nanotube, *Nature*, **393**, 49–52.

8. (2005). International Technology Roadmap for Semiconductors. Available: http://www.itrs.net/Links/2005ITRS/Home2005.htm.

9. Wei, B. Q., Vajtai, R., and Ajayan, P. M. (2001). Reliability and current carrying capacity of carbon nanotubes, *Appl. Phys. Lett.*, **79**, 1172–1174.

10. Raychowdhury A., and Roy, K. (2004). Modeling and analysis of carbon nanotube interconnects and their effectiveness for high speed VLSI design, *Proceedings of the 4th IEEE Conference on Nanotechnology*, pp. 608–610.

11. Naeemi, A., Sarvari, R., and Meindl, J. D. (2005). Performance comparison between carbon nanotube and copper interconnects for gigascale integration (GSI), *IEEE Electron Device Lett.*, **26**, 84–86.

12. Srivastava, N., and Banerjee, K. (2005). Performance analysis of carbon nanotube interconnects for VLSI applications, *Proceedings of the IEEE/ACM International Conference on Computer Aided Design*, 383–340.

13. Naeemi, A., and Meindl, J. D. (2007). Design and performance modeling for single-walled carbon nanotubes as local, semiglobal, and global interconnects in gigascale integrated systems, *IEEE Trans. Electron Devices*, **54**, 26–37.

14. Wong, H. S. P. (2002). Field effect transistors—from silicon MOSFETs to carbon nanotube FETs, *Proceedings of the 23rd International Conference on Microelectronics* (*MIEL 2002*), **1**, 103–107.

15. Wind, S. J., Appenzeller, J., Martel, R., Derycke, V., and Avouris, P. (2002). Vertical scaling of carbon nanotube field-effect transistors using top gate electrodes, *Appl. Phys. Lett.*, **80**, 3817–3819.

16. Guo, J., Datta, S., Lundstrom, M., Brink, M., McEuen, P., Javey, A., Hongjie, D., Hyoungsub, K., and P. McIntyre, (2002). Assessment of silicon MOS and carbon nanotube FET performance limits using a general theory of ballistic transistors, *IEDM Tech. Dig.*, 711–714.

17. Fuhrer, M. S. (2003) Single-walled carbon nanotubes for nanoelectronics, in *Advanced Semiconductor and Nano Technologies* (Part II) Chapter 6 (ed. Morkoc, H.), Elsevier Science & Technology.

18. DeHon, A., and Likharev, K. K. (2005). Hybrid CMOS/nanoelectronic digital circuits: devices, architectures, and design automation, *Proceedings of the 2005 IEEE/ACM International Conference on Computer-Aided Design*, pp. 375–382.

19. Bacon, R. (1960). Growth, structure, and properties of graphite whiskers, *J. Appl. Phys.*, **31**, 283–290.

20. Oberlin, A., Endo, M., and Koyama, T. (1976). Filamentous growth of carbon through benzene decomposition, *J. Cryst. Growth*, **32**, 335–349.

21. Kroto, H. W., Heath, J. R., O'Brien, S. C., Curl, R. F., and Smalley, R. E. (1985). C_{60}: Buckminsterfullerene, *Nature*, **318**, 162–163.

22. Iijima, S. (1991). Helical microtubules of graphitic carbon, *Nature*, **354**, 56–58.

23. Bethune, D. S., Kiang, C. H., de Vries, M. S., Gorman, G., Savoy, R., Vazquez, J., and Beyers, R. (1993). Cobalt-catalyzed growth of carbon nanotubes with single-atomic-layer walls, *Nature*, **363**, 605–607.

24. Iijima, S., and Ichihashi, T. (17 June 1993). Single shell carbon nanotubes of 1 nm diameter, *Nature*, **363**, 603–605.

25. Saito, R., Dresselhaus, G., and Dresselhaus, M. S. (1998) *Physical Properties of Carbon Nanotubes*, (Imperial College Press, London, UK).

26. Dai, H. (2002). Carbon nanotubes: Synthesis, integration, and properties, *Acc. Chem. Res.*, **35**, 1035–1044.

27. McEuen, P. L., Bockrath, M., Cobden, D. H., Yoon, Y.-G., and Louie, S. G. (1999). Disorder, pseudospins and backscattering in carbon nanotubes, *Phys. Rev. Lett.*, **83**, 5098–5101.

28. Dresselhaus, M. S., Dresselhaus, G., and Avouris, P. (2001). *Carbon Nanotube: Synthesis, Properties, Structure, and Applications*, (Springer Verlag).

29. (2010) Nanotube Modeler Available: http://jcrystal.com/products/wincnt/, 2010.

30. (2010) Carbon nanotube. Available: http://en.wikipedia.org/wiki/, 2010.

31. McEuen, P. L., Fuhrer, M. S., and Park, H. (2002). Single-walled carbon nanotube electronics, *IEEE Trans. Nanotechnol.*, **1**, 78–85.

32. Kasumov, A. Y., Khodos, I. I., Ajayan, P. M., and Colliex, C. (1996). Electrical resistance of a single carbon nanotube, *Europhys. Lett.*, **34**, 429–434.

33. Li, J., Ye, Q., Cassell, A., Ng, H. T., Stevens, R., Han, J., and Meyyappan, M. (2003). Bottom-up approach for carbon nanotube interconnects, *Appl. Phys. Lett.*, **82**, 2491–2493.

34. Nihei, M., Horibe, M., Kawabata, A., and Awano, Y. (2004). Carbon nanotube vias for future LSI interconnects, *Proceedings of the IEEE 2004 International Interconnect Technology Conference*, pp. 251–253.

35. Pennington, G., Akturk, A., and Goldsman, N. (2003). Electron mobility of a semiconducting carbon nanotube, *Proceedings of the International Semiconductor Device Research Symposium*, pp. 412–413.

36. Dürkop, T., Getty, S. A., Cobas, E., and Fuhrer, M. S. (2003). Extraordinary mobility in semiconducting carbon nanotubes, *Nano Lett.*, **4**, 35–39.

37. Yang, M. H., Teo, K. B. K., Gangloff, L., Milne, W. I., Hasko, D. G., Robert, Y., and Legagneux, P. (2006). Advantages of top-gate, high-κ dielectric carbon nanotube field-effect transistors, *App. Phys. Lett.*, **88**, 113507-1 to 13507-3.

38. Chen, Z., Appenzeller, J., Solomon, P. M., Lin, Y.-M., and Avouris, P. (2006). High performance carbon nanotube ring oscillator, *Proceedings of the 64th Device Research Conference*, pp. 171–172.

39. Hou, Y.-T., Li, M.-F., Low, T., and Dim-Lee, K. (2004). Metal gate work function engineering on gate leakage of MOSFETs, *IEEE Trans. Electron Devices*, **51**, 1783–1789.

40. Lu, C.-C., Chang-Liao, K.-S., Lu, C.-Y., Chang, S.-C., and Wang, T.-K. (2007). Leakage effect suppression in charge pumping measurement and stress-induced traps in high-κ Gated MOSFETs, *Proceedings of the International Semiconductor Device Research Symposium*, pp. 1–2.

41. Radosavljevic, M., Chu-Kung, B., Corcoran, S., Dewey, G., Hudait, M. K., Fastenau, J. M., Kavalieros, J., Liu, W. K., Lubyshev, D., Metz, M., Millard, K., Mukherjee, N., Rachmady, W., Shah, U., and Chau, R. (2009). Advanced high-κ gate dielectric for high-performance short-channel $In_{0.7}Ga_{0.3}As$ quantum well field effect transistors on silicon substrate for low power logic applications, *IEDM Tech. Dig.*, 13.1.1–13.1.4.

42. Ren, Z., Fischetti, M. V., Gusev, E. P., Cartier, E. A., and Chudzik, M. (2003). Inversion channel mobility in high-κ high performance MOSFETs, *IEDM Tech. Dig.*, 793–796.

43. Javey, A., Guo, J., Farmer, D. B., Wang, Q., Yenilmez, E., Gordon, R. G., Lundstrom, M., and Dai, H. (2004). Self-aligned ballistic molecular transistors and electrically parallel nanotube arrays, *Nano Lett.*, **4**, 1319–1322.

44. Javey, A., Guo, J., Paulsson, M., Wang, Q., Mann, D., Lundstrom, M., and Dai, H. (2004). High-field quasiballistic transport in short carbon nanotubes, *Phys. Rev. Lett.*, **92**, 106804.

45. Javey, A., Guo, J., Wang, Q., Lundstrom, M., and Dai, H. (2003). Ballistic carbon nanotube field-effect transistors, *Nature*, **424**, 654–657.

46. Javey, A., Kim, H., Brink, M., Wang, Q., Ural, A., Guo, J., McIntyre, P., McEuen, P., Lundstrom, M., and Dai, H. (2002). High-κ dielectrics for advanced carbon-nanotube transistors and logic gates, *Nat. Mater.*, **1**, 241–246.

47. Martel, R., Schmidt, T., Shea, H. R., Hertel, T., and Ph, A. (1998). Single- and multi-wall carbon nanotube field-effect transistors, *Appl. Phys. Lett.*, **73**, 2447–2449.

48. Radosavljevic, M., Appenzeller, J., Avouris, P., and Knoch, J. (2004). High performance of potassium n-doped carbon nanotube field-effect transistors, *Appl. Phys. Lett.*, **84**, 3693–3695.

49. Alam, K., and Lake, R. (2005). Performance of 2 nm gate length carbon nanotube field-effect transistors with source/drain underlaps, *Appl. Phys. Lett.*, **87**, 073104-1 to 073104-3.

50. Peng, H. B., Hughes, M. E., and Golovchenko, J. A. (2006). Room-temperature single charge sensitivity in carbon nanotube field-effect transistors, *Appl. Phys. Lett.*, **89**, 243502-1 to 243502-3.

51. Zhang, Z. Y., Wang, S., Ding, L., Liang, X. L., Xu, H. L., Shen, J., Chen, Q., Cui, R. L., Li, Y., and Peng, L. M. (2008). High-performance n-type carbon nanotube field-effect transistors with estimated sub-10-ps gate delay, *Appl. Phys. Lett.*, **92**, 133117-1 to 133117-3.

52. Aissa, B., and El, M. A. (2009). The channel length effect on the electrical performance of suspended-single-wall-carbon-nanotube-based field effect transistors, *Nanotechnology*, **20**, 175203-1 to 175203-2.

53. Ali Usmani, F., and Hasan, M. (2010). Carbon nanotube field effect transistors for high performance analog applications: An optimum design approach, *Microelectron. J.*, **41**, 395–402.

54. Derycke, V., Martel, R., Appenzeller, J., and Avouris, P. (2001). Carbon nanotube inter- and intramolecular logic gates, *Nano Lett.*, **1**, 453–456.

55. Martel, R., Derycke, V., Appenzeller, J., Wind, S., and Avouris, P. (2002). Carbon nanotube field-effect transistors and logic circuits, *Proceedings of the 39th Design Automation Conference*, pp. 94–98.

56. Zhang, Z., Liang, X., Wang, S., Yao, K., Hu, Y., Zhu, Y., Chen, Q., Zhou, W., Li, Y., Yao, Y., Zhang, J., and Peng, L.-M. (2007). Doping-free fabrication of carbon nanotube based ballistic CMOS devices and circuits, *Nano Lett.*, **7**, 3603–3607.

57. Kong, J., Yenilmez, E., Tombler, T. W., Kim, W., Dai, H., Laughlin, R. B., Liu, L., Jayanthi, C. S., and. Wu, S. Y. (2001). Quantum interference and ballistic transmission in nanotube electron waveguides, *Phys. Rev. Lett.*, **87**, 106801.

58. Niu, R., Zhang, J., Wang, Z., Liu, G., Zhang, G., Ding, X., and Sun, J. (2009). Mechanical strength lowering in submicron Cu thin films by moderate DC current, *Appl. Phys. A*, **97**, 369–374.

59. Kreupl, F., Graham, A. P., Duesberg, G. S., Steinhögl, W., Liebau, M., Unger, E., and Hönlein, W. (2002). Carbon nanotubes in interconnect applications, *Microelectron. Eng.*, **64**, 399–408.

60. Ngo, Q., Petranovic, D., Krishnan, S., Cassell, A. M., Ye, Q., Li, J., Meyyappan, M., and Yang, C. Y. (2004). Electron transport through metal-multiwall carbon nanotube interfaces, *IEEE Trans. Nanotechnol.*, **3**, 311–317.

61. Park, J.-Y., Rosenblatt, S., Yaish, Y., Sazonova, V., Ustunel, H., Braig, S., Arias, T. A., Brouwer, P. W., and McEuen, P. L. (2004). Electron–phonon scattering in metallic single-walled carbon nanotubes, *Nano Lett.*, **4**, 517–520.

62. Anantram, M. P., and Léonard, F. (2006). Physics of carbon nanotube electronic devices, *Rep. Prog. Phys.*, **69**, 507–561.

63. Anantram, M. P., and Govindan, T. R. (1998). Conductance of carbon nanotubes with disorder: A numerical study, *Phys. Rev. B*, **58**, 4882.

64. White, C. T., and Mintmire, J. W. (1998). Density of states reflects diameter in nanotubes, *Nature*, **394**, 29–30.

65. Liang, X., Wang, S., Wei, X., Ding, L., Zhu, Y., Zhang, Z., Chen, Q., Li, Y., Zhang, J., and Peng, L.-M. (2009). Towards entire-carbon-nanotube circuits: the fabrication of single-walled-carbon-nanotube field-effect transistors with local multiwalled-carbon-nanotube interconnects, *Adv. Mater.*, **21**, 1339–1343.

66. Nihey, F., Hongo, H., Ochiai, Y., Yudasaka, M., and Iijima, S. (2003). Carbon-nanotube field-effect transistors with very high intrinsic transconductance, *Jpn. J. Appl. Phys.*, **42**, L1288.

67. Park, J. W., Choi, J. B., and Yoo, K.-H. (30 Sept. 2002). Formation of a quantum dot in a single-walled carbon nanotube using the Al top-gates, *Appl. Phys. Lett.*, **81**, 2644–2646.

68. Nihey, F., Hongo, H., Yudasaka, M., and Iijima, S. (2002). A top-gate carbon-nanotube field-effect transistor with a titanium-dioxide insulator, *Jpn. J. Appl. Phys.*, **41**, L1049–L1051.

69. O'Connor, I., Liu, J., Gaffiot, F., Pregaldiny, F., Lallement, C., Maneux, C., Goguet, J., Fregonese, S., Zimmer, T., Anghel, L., Trong-Trinh, D., and Leveugle, R. (2007). CNTFET modeling and reconfigurable logic-circuit design, *IEEE Transactions on Circuits and Systems I:* Regular Papers, **54**, 2365–2379.

70. Guo, J., Koswatta, S. O., Neophytou, N., and Lundstrom, M. (2006). Carbon nanotube field-effect transistors, *Int. J. High Speed Electron. Syst.*, **16**, 897–912.

71. Natori, K., Kimura, Y., and Shimizu, T. (2005). Characteristics of a carbon nanotube field-effect transistor analyzed as a ballistic nanowire field-effect transistor, *J. Appl. Phys.*, **97**, 034306-1 to 034306-7.

72. Raychowdhury, A., Mukhopadhyay, S., and Roy, K. (2004). A circuit-compatible model of ballistic carbon nanotube field-effect transistors, *IEEE Trans. Comput.-Aided Des. Integrated Circuits Syst.*, **23**, 1411–1420.

73. Hazeghi, A., Krishnamohan, T., and Wong, H. S. P. (2007). Schottky-barrier carbon nanotube field-effect transistor modeling, *IEEE Trans. Electron Devices*, **54**, 439–445.

74. Hashempour, H., and Lombardi, F. (2007). Circuit-level modeling and detection of metallic carbon nanotube defects in carbon nanotube FETs, Proceedings Design, Automation & Test in Europe Conference & Exhibition (DATE), Nice, France, pp. 1–6.

75. Srivastava, A., Marulanda, J. M., Xu, Y., and Sharma, A. K. (2009). Current transport modeling of carbon nanotube field effect transistors, *Phys. Status Solidi (A)*, **206**, 1569–1578.

76. Mintmire, J. W., and White, C. T. (1998). Universal density of states for carbon nanotubes, *Phys. Rev. Lett.*, **81**, 2506–2509.

77. Tsividis, Y. (1999). *Operation and Modeling of the MOS Transistor.* (McGraw-Hill, Singapore).

78. Antognetti, P., and Massobrio, G. (1998). *Semiconductor Device Modeling with SPICE.* (McGraw-Hill, Singapore).

79. Marulanda, J. M., and Srivastava, A. (2007). Carrier density and effective mass calculations for carbon nanotubes, *Proceedings of the International Conference on Integrated Circuit Design & Technology (ICICDT)*, pp. 234–237.

80. Marulanda, J. M., Srivastava, A., and Nahar, R. K. (2005). Ultra-high frequency modeling of carbon nanotube field-effect transistors, *Proceedings of the IEEE 13th International Workshop on the Physics of Semiconductor Devices (IWPSD)*, New Delhi, India, pp. G-11.

81. Marulanda, J., Srivastava, A., and Sharma, A. K. (2007). Transfer characteristics and high frequency modeling of logic gates using carbon nanotube field effect transistors (CNT-FETs), *Proceedings of the 20th Annual Conference on Integrated Circuits and Systems Design*, Rio de Janeiro, Brazil, pp. 202–206.

82. Marulanda, J. M., Srivastava, A., and Sharma, A. K. (2008). Current transport modeling in carbon nanotube field effect transistors (CNT-FETs) and bio-sensing applications, *Proceedings of the SPIE Smart Structures and Materials & Nondestructive Evaluation and Health Monitoring: Nanosensors and Microsensors for Bio-System*, **6931**, 693108-1 to 693108-9.

83. Fisher, M. P. A., and Glazman, L. I. (1996). Transport in a one-dimensional Lüttinger liquid, in *Mesoscopic Electron Transport.*, **345,** (eds. Kowenhoven, L. et al.), Kluwer Academic Publishers, pp. 331–373.

84. Burke, P. J. (2003). An RF circuit model for carbon nanotubes, *IEEE Trans. Nanotechnol.*, **2**, 55–58.

85. Burke, P. J. (2002). Lüttinger Liquid theory as a model of the gigahertz electrical properties of carbon nanotubes, *IEEE Trans. Nanotechnol.*, **1**, 129–144.

86. Salahuddin, S., Lundstrom, M., and Datta, S. (2005). Transport effects on signal propagation in quantum wires, *IEEE Trans. Electron Devices*, **52**, 1734–1742.

87. Fetter, A. L. (1973). Electrodynamics of a layered electron gas. I. single layer, *Ann. Phys.*, **81**, 367–393.

88. Fetter, A. L. (1974). Electrodynamics of a layered electron gas. II. periodic array, *Ann. Phys.*, **88**, 1–25.

89. Maffucci, A., Miano, G., and Villone, F. (2008). A transmission line model for metallic carbon nanotube interconnects, *Int. J. Theory Appl.*, **36**, 31–51.

Chapter 2

Current Transport in Carbon Nanotubes

2.1 Introduction

Since the discovery of carbon nanotubes in 1991 [1], significant amount of research has been conducted to study its electronic properties [2–5]. Carbon nanotubes are being predicted to be the future material to substitute silicon used in CMOS technology at the end of Moore's law [6–8]. As it was explained in Chapter 1, carbon nanotubes are one-dimensional (1D) graphene sheets rolled into a tubular form [9]. Their electronic properties depend on its diameter and wrapping angle [2], which is represented by the indices (n,m) defined in the chiral vector characterizing each carbon nanotube. The electronic structure and electrical properties have been theoretically studied [3,10] based on the band theory of graphite [11,12] and have also been established experimentally [2]. The density of states has also been calculated [13,14] and is directly related to the chiral vector of the carbon nanotube [15].

Although applications of carbon nanotubes have increased over the past decade, very little work has been performed on modeling the carrier concentration. Recently, Raychowdhury et al. [16] have presented equations for the carrier concentration in carbon nanotubes in an attempt to calculate the inside charge. However, their model uses numerical curve-fitting techniques. In this chapter, we have used the density of states function of Mintmire and White

Carbon-Based Electronics: Transistors and Interconnects at the Nanoscale
Ashok Srivastava, Jose Mauricio Marulanda, Yao Xu, and Ashwani K. Sharma
Copyright © 2015 Pan Stanford Publishing Pte. Ltd.
ISBN 978-981-4613-10-1 (Hardcover), 978-981-4613-11-8 (eBook)
www.panstanford.com

[14] to derive analytical equations, which predict the carrier concentration in carbon nanotubes.

2.1.1 Energy Dispersion Relation

The energy dispersion relation in carbon nanotubes is calculated from the electronic structure of graphene [17,18]. The one-dimensional energy dispersion relation for single-walled carbon nanotubes is given by [19–21]

$$E_{1D}(k) = \pm V_{pp\pi}\left[1 + 4\cos\left(\frac{\sqrt{3}K_x}{2}a\right)\cos\left(\frac{K_y}{2}a\right) + 4\cos^2\left(\frac{K_y}{2}a\right)\right]^{1/2}, \quad (2.1)$$

where $V_{pp\pi}$ is the nearest neighbor overlap integral between C–C atoms used in tight binding calculations of the carbon nanotube. In the present research, we have used $V_{pp\pi}$ = 2.5 eV [17]. K_x and K_y are the wave vectors of carbon nanotubes [17,18], and they are given by

$$K_x = \frac{j2\pi\sqrt{3}a(n+m)C_h + a^3k(n^3-m^3)}{2C_h^3}$$

and

$$K_y = \frac{j2\pi a(n-m)+\sqrt{3}akC_h(n+m)}{2C_h^2},$$

where, C_h is the chiral vector given by $C_h = na_1 + ma_2$ and a_1 and a_2 are the unit vectors for the graphene hexagonal structure [2].

2.1.2 Density of States

Numerical techniques are needed to compute the density of states from Eq. (2.1) due to its complexity. However, an approximate density of states calculation has already been found for carbon nanotubes [14,22] and is described as follows:

$$D(E)dE = 2\sum_i^{AllBands} \frac{4}{\pi V_{pp\pi}a\sqrt{3}}\frac{E}{\sqrt{E^2-E_{cmin_i}^2}}dE, \quad (2.2)$$

where E_{cmin_i} is the minimum energy value for the given conduction band. E_{cmin_i} is found by determining the energy minimum value for the respective conduction band using Eq. (2.1). The first conduction band, E_{cmin}, can also be obtained from the following approximated equation [17]:

$$E_{cmin} = \frac{E_g}{2} = \frac{aV_{pp\pi}}{d\sqrt{3}},$$

(2.3)

where d is the diameter of the carbon nanotube and E_g is the energy bandgap.

2.2 Effective Mass

Given the complete description of the energy dispersion for carbon nanotubes, Eq. (2.1) can also be used to calculate the electron effective mass for each band. We can use the effective mass relationship in a semiconductor [23] for calculating the effective electron mass in a CNT (n,m) as follows:

$$m_i^* = \frac{\hbar^2}{\left(\dfrac{d^2E}{dk^2}\right)}$$

(2.4)

Table 2.1 summarizes the electron effective mass for various carbon nanotubes (n,m) calculated from Eqs. (2.4) and (2.1).

Table 2.1 Effective mass of electrons in carbon nanotubes

(n,m)	Effective mass of electrons (m^*)
(3,1)	$0.507\ m_0{}^\dagger$
(3,2)	$0.222\ m_0$
(4,2)	$0.271\ m_0$
(4,3)	$0.175\ m_0$
(5,0)	$0.408\ m_0$
(5,1)	$0.159\ m_0$
(5,3)	$0.189\ m_0$

Table 2.1 *(Continued)*

(n,m)	Effective mass of electrons (m*)
(6,1)	$0.255\,m_0$
(7,3)	$0.116\,m_0$
(9,2)	$0.099\,m_0$
(11,3)	$0.108\,m_0$

†m_0 is the mass of the electron (9.109×10^{-31} kg).

2.3 Carrier Concentration

The carrier concentration in a semiconductor is given by [23–25]

$$n_{\text{cnt}} = \int_{E_c}^{\infty} D(E)f(E)dE, \tag{2.5}$$

where $D(E)$ is the density of states, $f(E)$ is the Fermi level and E_c is the conduction band minimum value. Substituting Eq. (2.2) for the density of states in Eq. (2.5), we obtain the equation for carrier concentration given by

$$n_{\text{cnt}} = 2 \sum_{i}^{\text{AllBands}} \left[\frac{4}{\pi V_{\text{pp}\pi}a\sqrt{3}} \int_{E_{c_i}}^{\infty} E(E^2 - E_{c_i}^2)^{-1/2}(1 + e^{(E-E_F)/kT})^{-1}dE \right] \tag{2.6}$$

Equation (2.6) can be further simplified and expressed in the following form:

$$n_{\text{cnt}} = \frac{8}{\pi V_{\text{pp}\pi}a\sqrt{3}} \int_{0}^{\infty} (E' + E_c)(E'^2 + 2E_c E')^{-1/2}(1 + e^{(E'-E_F+E_c)/kT})^{-1}dE' \tag{2.7}$$

In deriving Eq. (2.7), the limits of integration in Eq. (2.6) have been changed by replacing the variable E with $(E_c + E')$. Furthermore, the summation has also been dropped as the Fermi function becomes negligible for conduction energy band minimums beyond the first band.

The integral in Eq. (2.7) is still very difficult to solve analytically; nevertheless, by putting $x = E/kT$ and $\eta = (E_F - E_c)/kT$, we can write Eq. (2.7) as

$$n_{cnt} = \frac{8\sqrt{kT}}{\pi V_{pp\pi} a\sqrt{3}} \int_0^\infty (kTx + E_c)\left[x(kTx + 2E_c)\right]^{-1/2} (1 + e^{x-\eta})^{-1} dx. \quad (2.8)$$

In addition, by defining $G(x) = (kTx + E_c)/x^{1/2}(kT_x + 2E_c)^{1/2}$ and $F(x) = 1/(1 + e^{x-\eta})$, Eq. (2.8) can be rewritten as

$$n_{cnt} = N_c \frac{1}{\sqrt{kT}} \int_0^\infty G(x)F(x)dx = N_c \frac{1}{\sqrt{kT}} \int_0^\infty \frac{G(x)}{1 + e^{x-\eta}} dx, \quad (2.9)$$

where

$$N_c = \frac{8kT}{\pi V_{pp\pi} a\sqrt{3}}.$$

The integral in Eq. (2.9) is similar to the Fermi integral in [24] and can be approximately integrated under two limits described as follows [23,24].

2.3.1   Limit 1: $\eta \ll -1$

Under this limit, it can be shown that the function $G(x)$ retains a constant value of $\sqrt{kT}$ for $x \gg E_c/kT$. Therefore, the upper limit of the integral can be replaced from infinity to $6E_c/kT$ since for values beyond this limit, $6E_c/kT$, the exponential function $F(x)$ approaches zero and the integral becomes negligible. Under these conditions, it is possible to write

$$n_{cnt} = N_c I e^{(E_F - E_c)/kT}, \quad (2.10)$$

where

$$I = \frac{1}{\sqrt{kT}} \int_0^{6E_c/kT} \frac{(kTx + E_c)}{x^{1/2}(kTx + 2E_c)^{1/2}} e^{-x} dx.$$

In Eq. (2.10), the integral, I, has no definite solution. However, an approximate solution has been found in the work of Marulanda and Srivastava [26] by representing the exponential function, e^{-x}, with a series of polynomial functions using the Taylor series expansion approximation [27,28] around a variable A. The solution can be expressed as

$$
I = \sum_{A=0}^{\mathrm{int}(6(E_c/kT))} \left\{ e^{-A} \sqrt{x\left(x+2\frac{E_c}{kT}\right)} \left[1 - \frac{1}{2}x + A + \frac{1}{2}\frac{E_c}{kT} + \frac{1}{6}x^2 \right.\right.
$$

$$
-\frac{1}{6}\frac{E_c}{kT}x - \frac{1}{2}Ax + \frac{1}{2}\left(\frac{E_c}{kT}\right)^2 + \frac{1}{2}A\frac{E_c}{kT}
$$

$$
+\frac{1}{2}A^2 - \frac{1}{4}A^2 x + \frac{5}{16}\left(\frac{E_c}{kT}\right)^3 + \frac{1}{6}A^3
$$

$$
-\frac{1}{24}x^3 + \frac{1}{24}\frac{E_c}{kT}x^2 - \frac{5}{48}\left(\frac{E_c}{kT}\right)^2 x
$$

$$
+\frac{1}{2}A\left(\frac{E_c}{kT}\right)^2 + \frac{1}{2}A^2\frac{E_c}{kT} + \frac{1}{6}Ax^2 - \frac{1}{6}A\frac{E_c}{kT}x \right]
$$

$$
+\frac{1}{2}\left(\frac{E_c}{kT}\right)^2 \left[1 + A + \frac{E_c}{kT} + \frac{1}{2}A^2 - A\frac{E_c}{kT} - \frac{5}{8}\left(\frac{E_c}{kT}\right)^2 \right].
$$

$$
\left.\ln\left[\frac{E_c}{kT} + \sqrt{kT} + \sqrt{kTx^2 + 2xE_c}\right]\Bigg|_{x=A}^{x=A+1} \right\}.
$$

$$\tag{2.11}$$

2.3.2 Limit 2: $\eta \gg 1$

Under this limit, the exponential function $F(x)$ of Eq. (2.9) can be approximated as $F(x) = 1$, which will remain true as long as $x < \eta$. This approximation fails for $x > \eta$, which is the case of Eq. (2.9), since the upper limit of integration is infinite. However, for $x > \eta$, the exponential term in $F(x)$, $e^{x-\eta}$, becomes very large causing $F(x)$ to approach zero. Therefore, $F(x)$ can be considered negligible for $x > \eta$. This latter approximation allows us to change the upper limit of integration since $F(x)$ vanishes for $x > \eta$. Thus, the upper limit of Eq. (2.9) can be replaced with $x = \eta$.

Under these conditions Eq. (2.9) becomes

$$
n_{cnt} = N_c \frac{(E_F^2 - E_c^2)^{1/2}}{kT}. \tag{2.12}
$$

For intrinsic level calculations and normal doping, Limit 1 is useful (Eq. 2.10); Limit 2 (Eq. 2.12) becomes important for heavy doping. By setting $E_F = E_i$, where E_i is the intrinsic energy level and lies in the middle of the bandgap, we can obtain the intrinsic carrier concentration, $n_{cnt,i}$ and is given by

$$n_{cnt,i} = N_c I e^{-E_c/kT}. \tag{2.13}$$

Table 2.2 shows the intrinsic carrier concentration for different carbon nanotubes (n,m) obtained from Eq. (2.13) at room temperature (300 K). An effective carbon nanotube wall thickness of 0.617 Å [29] has been used in the calculations of Table 2.2. The fourth column in Table 2.2 summarizes intrinsic carrier concentration in carbon nanotube per unit volume for specific chiral vector (n,m), energy bandgap (E_g) and diameter (d). Since single-walled carbon nanotubes can be considered true quasi-one-dimensional conductors, the fifth column in Table 2.2 summarizes intrinsic carrier concentration in carbon nanotubes per unit length. It is noticed from Table 2.2 that a carbon nanotube as a quasi-one-dimensional behaves as an insulator for certain combinations of chiral vectors, energy bandgap and diameter. For other combinations, the carbon nanotube behaves as a conductor.

Table 2.2   Intrinsic carrier concentration in carbon nanotubes

(n,m)	d (nm)	E_g (eV)	$n_{cnt,i}$ (cm^{-3})	$n_{cnt,i}$ (cm^{-1})
(3,2)	0.344	2.089	7.042×10^4	0
(4,2)	0.417	1.721	6.303×10^6	0
(4,3)	0.480	1.497	1.239×10^9	0
(5,0)	0.394	1.821	2.401×10^5	0
(5,1)	0.439	1.635	2.677×10^8	0
(5,3)	0.552	1.301	1.791×10^{10}	0
(6,1)	0.517	1.389	1.656×10^9	0
(7,3)	0.701	1.024	5.238×10^{12}	0
(9,2)	0.800	0.897	4.911×10^{13}	0
(9,8)	1.161	0.618	7.748×10^{15}	1.70×10^1
(10,8)	1.232	0.583	1.404×10^{16}	3.40×10^1
(10,9)	1.298	0.553	2.314×10^{16}	5.80×10^1

Table 2.2 (*Continued*)

(n,m)	d (nm)	E_g (eV)	$n_{cnt,i}$ (cm^{-3})	$n_{cnt,i}$ (cm^{-1})
(11,3)	1.007	0.713	6.034×10^{14}	3.00×10^0
(11,6)	1.177	0.610	8.929×10^{15}	2.00×10^1
(11,10)	1.434	0.500	5.530×10^{16}	15.30×10^1
(12,8)	1.375	0.522	3.864×10^{16}	10.30×10^1
(14,13)	1.844	0.389	3.289×10^{17}	11.76×10^2
(20,19)	2.663	0.270	1.958×10^{18}	10.11×10^3
(21,19)	2.732	0.263	2.153×10^{18}	11.40×10^3
(40,38)	5.326	0.135	9.924×10^{18}	10.25×10^4

Equations (2.10) and (2.12) can also be used to study the effect of temperature on the intrinsic carrier concentration of carbon nanotubes. Figure 2.1 shows the dependence of intrinsic carrier concentration in carbon nanotubes obtained from Eq. (2.13) on temperature for three different chiral vectors.

In Fig. 2.1, we can clearly see the strong exponential dependence of the intrinsic carrier concentration on temperature for an intrinsic carbon nanotube with chiral vectors (4,2), (4,3), and (7,3).

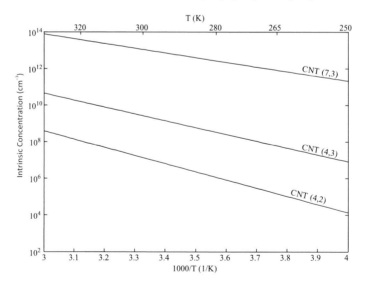

Figure 2.1 Plot of the intrinsic carrier dependence on temperature for a carbon nanotube with chiral vectors (4,2), (4,3) and (7,3).

The carrier concentration in a doped carbon nanotube can be obtained by adding an impurity concentration in Eq. (2.13) and is given by

$$n_{cnt} = n_{cnt,i} + N,$$ (2.14)

where N is the ionized impurity concentration in a carbon nanotube.

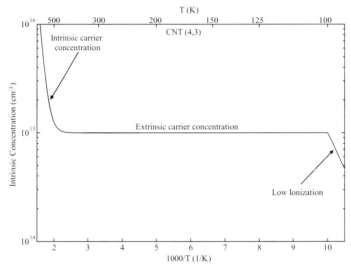

Figure 2.2 Plot of the carrier concentration dependence on temperature for a carbon nanotube with a chiral vector (4,3) and a doping concentration of 10^{15} donor atoms. Note: For high temperatures, the intrinsic concentration dominates and for low temperatures (below 100 K), the concentration decreases due to the incapability of the donor atoms to become ionized.

Assuming we have a carbon nanotube with a chiral vector (4,3) doped with 10^{15} donors, Fig. 2.2 shows the effect of temperature obtained from Eq. (2.14) and can be explained as follows: For low temperatures below 100K, there is low ionization and therefore low carrier concentration is observed (a value of 100 K for the ionization temperature of CNTs has been assumed from that of silicon [23]). As the temperature is increased, more and more electrons are available in the conduction band, and at about 100 K ($1000/T = 10$), all donor atoms are ionized and the carrier concentration equals that of the donor atoms. This temperature range, ~100 to 500 K, is the ionization region for carbon nanotubes. Beyond ~500 K, the intrinsic carrier concentration dominates as predicted by the exponential dependence shown in Fig. 2.1. The behavior is similar to carrier concentration dependence on temperature in silicon [23].

The effect of donor atoms on the Fermi energy level of carbon nanotubes has been studied using Eqs. (2.10) and (2.12). In Fig. 2.3, we can see the variation of the energy separation $(E_{co} - E_F)$ under normal and heavy doping for two carbon nanotubes with chiral vectors (4,2) and (10,0).

In Fig. 2.3, we can clearly see the exponential dependence of the Fermi energy level on the doping concentration. This is confirmed by the linear dependence observed in the plot of the energy separation $(E_F - E_C)$ versus ln (*doping concentration*) for carbon nanotubes.

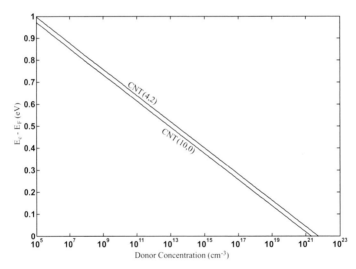

Figure 2.3 Plot of the energy separation $(E_F - E_C)$ versus doping concentration for two carbon nanotubes with chiral vectors (4,2) and (10,0).

2.4 Summary

The effective mass and carrier concentration in carbon nanotubes have been theoretically studied. A concise derivation of the carrier density under two limiting cases has been presented including analytical solutions for calculating the intrinsic carrier concentrations and effective mass in carbon nanotubes. Temperature dependences of the carrier concentration and energy band structure have been established analytically. The calculations obtained provide useful understanding of the conductivity in carbon nanotubes and electrical modeling. They represent an important contribution to this field of research; especially, when dealing with impurities, doping concentrations and its effects on the electronic band structure of hexagonal crystal lattice materials. In addition, looking ahead into the application of carbon nanotubes in field effect transistors (CNT-FETs), the analytical equations presented can also be used in calculating and deriving current transport models for these transistors. The analysis can be extended to bundles of carbon nanotubes comparing to a bulk material for interconnect applications.

References

1. Iijima, S. (1991). Helical microtubules of graphitic carbon, *Nature*, **354**, 56–58.

2. Wildoer, J., Venema, L., Rinzler, A., Smalley, R., and Dekker, C. (1998). Electronic structure of atomically resolved carbon nanotubes, *Nature*, **391**, 59–62.

3. Hamada, N., Sawada, S. I., and Oshiyama, A. (1992). New one dimensional conductors: Graphite microtubules, *Phys. Rev. Lett.*, **68**, 1579–1581.

4. Mintmire, J. W., Dunlap, B. I., and White, C. T. (1992). Are fullerene tubules metallic?, *Phys. Rev. Lett.*, **68**, 631–634.

5. Saito, R., Fujita, M., Dresselhaus, G., and Dresselhaus, M. S. (1992). Electronic structure of chiral graphene tubules, *Appl. Phys. Lett.*, **60**, 2204–2206.

6. Wong, H. S. P. (2002). Field effect transistors: From silicon MOSFETs to carbon nanotube FETs, *Proc. 23th International Conference on Microelectronics*, (*MIEL*), pp. 103–107.

7. Guo, J., Datta, S., Lundstrom, M., Brink, M., McEuen, P., Javey, A., Dai, H., Kim, H., and McIntyre, P. (2002). Assessment of silicon MOS carbon nanotube FET performance limits using a general theory of ballistic transistors, *IEDM Tech. Dig.*, 711–714.

8. Wind, S. J., Appenzeller, J., Martel, R., Derycke, V., and Avouris, P. (2002). Vertical scaling of carbon nanotube field-effect transistors using top gate electrodes, *Appl. Phys. Lett.*, **80**, 3817–3819.

9. Martel, R., Schmidt, T., Shea, H. R., Hertel, T., and Avouris, P. (1998). Single and multi-wall carbon nanotube field effect transistors, *Appl. Phys. Lett.*, **73**, 2447–2449.

10. Avouris, P. (2002). Molecular electronics with carbon nanotubes, *Acc. Chem. Res.*, **35**, 1026–1034.

11. Wallace, P. R. (1947). The band theory of graphite, *Phys. Rev. Lett.*, **71**, 622–634.

12. Saito, R., Fujita, M., Dresselhaus, G., and Dresselhaus, M. S. (1992). Electronic band structure of graphene tubules based on C_{60}, *Phys. Rev. Lett. B*, **46**, 1804–1811.

13. Martel, R., Derycke, V., Appenzeller, J., Wind, S., and Avouris, P. (2002). Carbon nanotube field effect transistors and logic circuits, *Proc. 39th Design Automation Conference*, pp. 94–98.

14. Mintmire J. W., and White, C. T. (1998). Universal density of states for carbon nanotubes, *Phys. Rev. Lett.*, **81**, 2506–2509.

15. White C. T, and Mintmire, J. W. (1998). Density of states reflects diameter in nanotubes, *Nature*, **394**, 29–30.

16. Raychowdhury, A., Mukhopadhyay, S., and Roy, K. (2004). A circuit-compatible model of ballistic carbon nanotube field effect transistors, *IEEE Trans. Comput.-Aided Des. Integrated Circ. Syst.*, **23**, 1411–1420.

17. Dresselhaus, M. S., Dresselhaus, G., and Avouris, P. (2001). *Carbon Nanotube: Synthesis, Properties, Structure, and Applications* (Springer-Verlag, Berlin Heidelberg New York).

18. Saito, R., Dresselhaus, M. S., and Dresselhaus, G. (1998). *Physical Properties of Carbon Nanotubes* (Imperial College Press, London, U.K.).

19. Loiseau, A., Launois, P., Petit, P., Roche, S., and Salvetat, J. P. (2006). *Understanding Carbon Nanotubes* (Springer-Verlag, Berlin Heidelberg).

20. Tanaka, K., Yamabe, T., and Fukui, K. (199). *The Science and Technology of Carbon Nanotubes* (Elsevier, Amsterdam, the Netherlands).

21. Saito, R., Dresselhaus, G., and Dresselhaus, M. S. (2000). Trigonal warping effect of carbon nanotubes, *Phys. Rev. Lett. B*, **61**, 2981–2990.

22. Guo, J., Lundstrom, M., and Datta, S. (2002). Performance projections for ballistic carbon nanotube field-effect transistors, *Appl. Phys. Lett.*, **80**, 3192–31942.

23. Streetman, B. G. (2000). *Solid State Electronic Devices*, 5th ed. India: Prentice Hall, India).

24. Shur, M. (1993). *Physics of Semiconductor Devices* (Prentice Hall, U.S.A.).

25. McKelvey, J. P. (1993). *Solid States Physics for Engineering and Material Sciences* (Krieger Publishing Company, Florida, U.S.A.).

26. Marulanda, J. M., and Srivastava, A. (2008). Carrier density and effective mass calculations in carbon nanotubes, *Phys. Status Solidi* (*B*), **245**, 2558–2562.

27. Johnson, D. E., and Johnson, J. R. (1965). *Mathematical Methods in Engineering and Physics* (The Ronald Press Company, New York, U.S.A.).

28. Stewart, J. (1965). *Calculus Early Transcendentals*, 3th ed. (Brooks/ Cole Publishing Company, U.S.A.).

29. Vodenitcharova, T., and Zhang, L. C. (2003). Effective wall thickness of a single-walled carbon nanotube, *Phys. Rev. Lett. B*, **68**, 156401, 2003.

Chapter 3

Current Transport in CNT Field-Effect Transistors

3.1 Introduction

Carbon nanotubes (CNTs), discovered in 1991 [1], have been a subject of intensive research for a wide range of applications in areas of chemical and biological sensing [2–4], emerging nanometer-sized field-effect transistors for very large scale integration and interconnects [5–10]. Carbon nanotubes are one-dimensional (1D) graphene sheets rolled into a tubular structure of nanometer size [11,12]. Their properties depend on the diameter and wrapping angle determined by the chiral vector, which is characterized by the indices (n,m) of the graphene [12]. Carbon nanotubes have been classified in two categories: single-walled carbon nanotubes (SWCNTs), which consist of a single layer of graphene, and multi-walled carbon nanotubes (MWCNTs), which consist of many stacked layers of graphene [11]. Carbon nanotube field-effect transistor (CNT-FET) has been identified as one of the promising candidates substituting shrinking CMOS technology at the end of the Moore's law [13–15].

The structure of a CNT-FET is similar to the structure of a typical MOSFET [13,16], where an SWCNT forms the channel between two electrodes, which work as the source and drain of the transistor. The structure is built on top of an insulating layer and

Carbon-Based Electronics: Transistors and Interconnects at the Nanoscale
Ashok Srivastava, Jose Mauricio Marulanda, Yao Xu, and Ashwani K. Sharma
Copyright © 2015 Pan Stanford Publishing Pte. Ltd.
ISBN 978-981-4613-10-1 (Hardcover), 978-981-4613-11-8 (eBook)
www.panstanford.com

a substrate, which works as the back gate [6,17]. Both the n- and p-type CNT-FETs have been fabricated in the past decade [18,19] and multistage complementary logic gates have been demonstrated [20–24].

With the advancement in fabrication technology of CNT-FETs, efforts have also been made in modeling of the transport behavior and models have been developed for the design of CNT-FETs based logic circuits [25–30]. CNT-FETs have also been modeled for the high frequency behavior [31–34]. In this chapter, we have attempted to develop analytical models characterizing the current transport in CNT-FETs for the analysis and design of integrated circuits. We have used these models for generating voltage transfer characteristics of basic logic gates such as the inverter, NAND, and NOR and shown the variation of the chiral vectors (*n,m*) of carbon nanotubes on their voltage transfer characteristics [35].

In Section 3.2, the current transport equation of a CNT-FET is obtained by relating the carbon nanotube potential to the terminal voltages. The charge inside the carbon nanotube is described from the electronic structure of the carbon nanotube. A model for the carbon nanotube potential is then derived and the current transport equation is obtained. In Section 3.3, voltage transfer characteristics of logic gates are presented using complementary CNT-FETs, followed by conclusion in Section 3.4.

3.2 Current Transport Modeling

Figure 3.1a shows the basic cross section of a CNT-FET, including the charge distributions. Figure 3.1b shows the corresponding potential distributions between the gate and the substrate. In Fig. 3.1a, charge distributions are explained as follows: the charge on the gate, Q_g, the charges in the oxide layers, Q_{01} and Q_{02}, the charge inside the CNT, Q_{cnt}, and the charge in the substrate, Q_{subs}. In Fig. 3.1b, six different potential distributions are shown, which are also described as follows. The voltage between the gate and the substrate (back gate) is V_{gb}, the potential drop across the oxides are ψ_{ox1} and ψ_{ox2}, the surface potential in the substrate with respect to the back gate is ψ_{subs}, the potential across the CNT is ψ_{cnt} and the work function difference between the gate and the substrate materials is ϕ_{ms}.

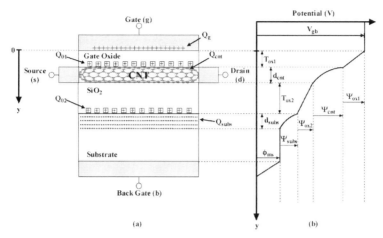

Figure 3.1   (a) Plot of the charges from the gate to the substrate and (b) plot of the potential distribution from the gate to the substrate in a CNT-FET.

Using Kirchoff's voltage law, the potential balance, and charge neutrality condition, we can write

$$V_{gb} = \phi_{ms} + \psi_{subs} + \psi_{ox2} + \psi_{cnt} + \psi_{ox1} \tag{3.1}$$

$$Q'_g + Q'_{01} + Q'_{cnT} + Q'_{02} + Q'_{subs} = 0, \tag{3.2}$$

The prime in Eq. (3.2) denotes the charge per unit area. In Eq. (3.1), ϕ_{ms} is divided in two parts and expressed as follows:

$$\phi_{ms} = \phi_{mc} + \phi_{cs}, \tag{3.3}$$

where ϕ_{mc} and ϕ_{cs} are the work function differences between the metal gate and carbon nanotube materials and the carbon nanotube and substrate materials, respectively.

Combining potentials ψ_{cnt}, ψ_{ox2}, ϕ_{ms}, and ψ_{subs} into a single potential, $\psi_{cnt,s}$, Eq. (3.1) can be re-written as follows:

$$V_{gb} = \phi_{mc} + \psi_{ox1} + \psi_{cnt,s}, \tag{3.4}$$

where $\psi_{cnt,s}$ describes the surface potential at the interface of the gate oxide and the carbon nanotube with respect to the back gate.

The electric field in terms of the charge distribution can be written from Maxwell's third equation [36] and is given by

$$\nabla(\varepsilon E) = \rho_\mathrm{v}, \tag{3.5}$$

where ε is the permittivity of the material, E is the electric field, and ρ_v is the charge per unit volume. Assuming that the electric field in Fig. 3.1a is constant through the gate oxide region and at the bottom edge of the carbon nanotube, Eq. (3.5) can be integrated between the gate and the bottom edge of the carbon nanotube, that is in Fig. 3.1a from $y = 0$ to $y = T_\mathrm{ox1} + d_\mathrm{cnt}$, to obtain the following Eq. (3.6), where d_cnt is the diameter of the carbon nanotube. In this derivation, we have assumed that the carbon nanotube has a relative permittivity, ε_cnt. We can write Eq. (3.6) as follows:

$$\varepsilon_\mathrm{cnt} E_\mathrm{cnt} - \varepsilon_\mathrm{ox1} E_\mathrm{ox1} = Q'_{01} + Q'_\mathrm{cnt}, \tag{3.6}$$

where E_cnt and E_ox1 are the electric fields along the y-axis across the carbon nanotube and gate oxide, respectively.

In a typical top-gated CNT-FET, the substrate oxide thickness, T_ox2, is much greater than the gate oxide thickness, T_ox1 [17]. We can then assume that any charge applied at the gate is compensated only by an induced charge in the carbon nanotube as follows:

$$\Delta Q'_\mathrm{g} = \Delta Q'_\mathrm{cnt} \tag{3.7}$$

Furthermore, there is a specific gate voltage, called the flat band voltage, V_fb [37], when applied at the gate with respect to the back gate; it compensates for the band bending at the gate oxide and carbon nanotube interface. Under this flat band condition, the electric field at the bottom edge of the carbon nanotube, E_cnt in Eq. (3.6) can be neglected. Under this assumption and replacing E_ox1 with the potential gradient $-\dfrac{d\psi_\mathrm{ox1}(y)}{dy}$ in Eq. (3.6) we obtain

$$\varepsilon_\mathrm{ox1} \frac{d\psi_\mathrm{ox1}(y)}{dy} = Q'_{01} + Q'_\mathrm{cn} \tag{3.8}$$

By integrating Eq. (3.8) through the gate oxide region, which is in Fig. 3.1a from $y = 0$ to $y = T_\mathrm{ox1}$, we can write an expression for the gate oxide potential as follows:

$$\psi_{\text{ox1}} = -\frac{Q'_{01} + Q'_{\text{cnt}}}{C'_{\text{ox1}}}. \tag{3.9}$$

In Eq. (3.9), C'_{ox1} is the gate oxide capacitance per unit area. Substituting Eq. (3.9) in Eq. (3.4), we obtain an expression for the gate voltage given by

$$V_{\text{gb}} = \psi_{\text{cnt,s}} - \frac{Q_{\text{cnt}}}{C_{\text{ox1}}} + \phi_{\text{mc}} - \frac{Q_{01}}{C_{\text{ox1}}}. \tag{3.10}$$

In Eq. (3.10), Q_{cnt}, Q_{01}, and C_{ox1} are the total charges and capacitance, respectively, which are obtained by multiplying Q'_{01}, Q'_{cnt}, and Q'_{02} with their respective areas. C_{ox1}, the capacitance between the gate and the carbon nanotube, can be redefined by considering the carbon nanotube to be a line of charge and the gate to be a planar conducting plate. Therefore, the gate oxide capacitance of thickness T_{ox1} of a carbon nanotube of length L and radius r is given by [38,39]

$$C_{\text{ox1}} = \frac{2\pi\varepsilon_{\text{ox1}} L}{\ln\left(\dfrac{T_{\text{ox1}} + r + \sqrt{T_{\text{ox1}}^2 + 2T_{\text{ox1}} r}}{r}\right)}. \tag{3.11}$$

In Eq. (3.10), the flat band voltage, V_{fb}, is

$$V_{\text{fb}} = \phi_{\text{mc}} - \frac{Q_{01}}{C_{\text{ox1}}}. \tag{3.12}$$

The gate voltage, V_{gb}, in Eq. (3.10) after combining with the Eq. (3.12) can be expressed as follows:

$$V_{\text{gb}} = \psi_{\text{cnt,s}} - \frac{Q_{\text{cnt}}}{C_{\text{ox1}}} + V_{\text{fb}}. \tag{3.13}$$

In Eq. (3.13), $\psi_{\text{cnt,s}}$ can be explained as a control potential in the carbon nanotube in charge of shifting the energy band at the interface of the gate oxide and carbon nanotube. As the gate voltage increases, we start seeing a voltage drop across the gate oxide modeled as Q_{cnt}/C_{ox1}. Equation (3.13) can be used to define the threshold voltage.

In Eq. (3.13), the charge inside the carbon nanotube can be calculated using the relation $|Q_{cnt}| = qn_{cnt}L$, where n_{cnt} is the carrier concentration per unit length inside the carbon nanotube, which we have derived in our earlier work [40,41] and is given by

$$n_{cnt}(\eta) = \frac{8\sqrt{kT}}{\pi\sqrt{3}V_{pp\pi}a} \int_0^\infty \frac{kTx + E_c}{\sqrt{x(kTx + 2E_c)}} (1 + e^{x-\eta})^{-1} dx, \tag{3.14}$$

where a is the lattice constant, $V_{pp\pi}$ is the energy transfer integral [11,42], k is Boltzmann's constant, and T is the temperature. The parameter, η is given by

$$\eta = \frac{E_F - E_c}{kT}, \tag{3.15}$$

where E_c is the conduction band minima. It is found from the energy dispersion relation of the carbon nanotubes and is expressed as follows [11,41]:

$$E_{1D}(k) = \pm V_{pp\pi} \left[1 + 4\cos\left(\frac{\sqrt{3}K_x}{2}a\right)\cos\left(\frac{K_y}{2}a\right) + 4\cos^2\left(\frac{K_y}{2}a\right) \right]^{1/2}, \tag{3.16}$$

where K_x and K_y are the wave vectors of a one-dimensional (1D) carbon nanotube.

Fregonese et al., [43] have attempted in finding an analytical solution for the Eq. (3.14). We have also reported in [40,41] a simple analytical solution under two limiting cases. Under these limiting conditions, we can express the charge inside the carbon nanotube as in [44], which is as follows:

Limit 1, $\eta << -1$:

$$|Q_{cnt}| = qLN_c Ie\frac{E_F - E_c}{kT}, \tag{3.17}$$

Limit 2, $\eta >> 1$:

$$|Q_{cnt}| = qLN_c \frac{\sqrt{E_F^2 - E_c^2}}{kT}, \tag{3.18}$$

where

$$I = \sum_{A=0}^{\mathrm{int}\left(6\frac{E_c}{kT}\right)} \left\{ e^{-A} \sqrt{x\left(x + 2\frac{E_c}{kT}\right)} \left[1 - \frac{1}{2}x + A + \frac{1}{2}\frac{E_c}{kT} + \frac{1}{6}x^2 \right.\right.$$

$$- \frac{1}{6}\frac{E_c}{kT}x - \frac{1}{2}Ax + \frac{1}{2}\left(\frac{E_c}{kT}\right)^2 + \frac{1}{2}A\frac{E_c}{kT}$$

$$+ \frac{1}{2}A^2 - \frac{1}{4}A^2x + \frac{5}{16}\left(\frac{E_c}{kT}\right)^3 + \frac{1}{6}A$$

$$- \frac{1}{24}x^3 - \frac{1}{24}\frac{E_c}{kT}x^2 - \frac{5}{48}\left(\frac{E_c}{kT}\right)^2 x$$

$$\left. + \frac{1}{2}A\left(\frac{E_c}{kT}\right)^2 + \frac{1}{2}A^2\frac{E_c}{kT} + \frac{1}{6}Ax^2 - \frac{1}{6}A\frac{E_c}{kT}x \right]$$

$$+ \frac{1}{2}\left(\frac{E_c}{kT}\right)^2 \left[1 + A + \frac{E_c}{kT} + \frac{1}{2}A^2 - A\frac{E_c}{kT} - \frac{5}{8}\left(\frac{E_c}{kT}\right)^2 \right].$$

$$\left. \ln\left[\frac{E_c}{kT} + \sqrt{kT} + \sqrt{kTx^2 + 2xE_c} \right]\right\}_{x=A}^{x=A+1}$$

(3.19)

and

$$N_c = \frac{8kT}{\pi\sqrt{3}V_{pp\pi}a}.$$

(3.20)

It should be noted that the solution of Eqs. (3.17) and (3.18) considers only the charge in the first sub-band.

In order to effectively describe the potential inside the carbon nanotube, we need to determine the energy separation between the Fermi level and the conduction band at the interface of the gate oxide and carbon nanotube. Assuming a flat band energy level in the carbon nanotube, we can say that the Fermi level and the conduction band at the interface of the gate oxide and carbon nanotube will be shifted by an amount determined by qV_{cb} and $q\psi_{cnt,s}$, respectively, where V_{cb} is the induced potential between the carbon nanotube and the substrate due to the drain and source terminal voltages. The potential, V_{cb} varies from V_{sb} (source to back gate potential) to V_{db} (drain to back gate potential). Figure 3.2

shows the energy band diagram from gate to substrate of a CNT-FET when (a) $V_{gb} = |V_{fb}|$ and (b) $V_{gb} > 0$. In Fig. 3.2a, we have used hafnium dioxide (HfO$_2$) as the gate oxide, which is a high k-dielectric insulator material. Furthermore, ϕ_0 in Fig. 3.2a is the carbon nanotube surface potential, $\psi_{cnt,s}$ when $V_{gb} = V_{fb}$ and $\psi_{cnt} = 0$. ϕ_{ms} is then given by

$$\phi_0 = \psi_{ox2} + \psi_{subs} + \phi_{cs} \tag{3.21}$$

(a)
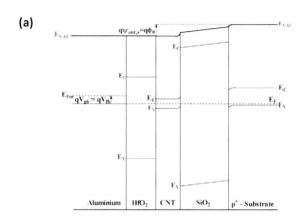

(b)
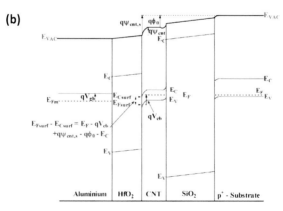

Figure 3.2 Energy band diagram of a two terminal CNT-FET for (a) $V_{gb} = V_{fb}$ and (b) $V_{gb} > 0$. Note: HfO$_2$ is the high k-dielectric hafnium oxide.

As it was previously done, by integrating Eq. (3.5) from $y = T_{ox1} + d_{cnt}$ to $y = T_{ox1} + d_{cnt} + T_{ox2} + d_{subs}$, where d_{subs} is the thickness of Q_{subs} in the substrate as shown in Fig. 3.1a, we can write

$$\varepsilon_s E_s - \varepsilon_{ox2} E_{ox2} = Q'_{02} + Q'_{subs} \tag{3.22}$$

where ε_s is the permittivity of the substrate material. Assuming the electric field, E_s is negligible deep inside the substrate and replacing E_{ox2} with the potential gradient $-\dfrac{d\psi_{ox2}(y)}{dy}$ in Eq. (3.22) we have

$$\varepsilon_{ox2} \frac{d\psi_{ox2}(y)}{dy} = Q'_{02} + Q'_{subs} \tag{3.23}$$

By integrating Eq. (3.23) from $y = T_{ox1} + d_{cnt}$ to $y = T_{ox1} + d_{cnt} + T_{ox2}$ in Fig. 3.1a we can obtain an expression for the second oxide potential as

$$\psi_{ox2} = -\frac{Q'_{02} + Q'_{subs}}{C'_{ox2}} \tag{3.24}$$

The total capacitance C_{ox2}, which is the capacitance between the carbon nanotube and the substrate and is obtained as follows [38,39]:

$$C_{ox2} = \frac{2\pi \varepsilon_{ox2} L}{\ln\left(\dfrac{T_{ox2} + r + \sqrt{T_{ox2}^2 + 2T_{ox2}r}}{r}\right)}. \tag{3.25}$$

Assuming Q_{subs} and ψ_{subs} to be small compared to Q_{02} and $\phi_{cs} + \psi_{ox2}$, we can rewrite Eq. (3.21) for ϕ_{ms} as follows:

$$\phi_0 = \phi_{cs} - \frac{Q_{02}}{C_{ox2}} \tag{3.26}$$

We are interested in the separation between the Fermi level and conduction band at the interface of the gate oxide and carbon nanotube. Following Fig. 3.2a we can observe that the conduction band of the carbon nanotube is shifted by an amount of $\psi_{cnt,s} - \phi_{ms}$ expressed as [25,37]

$$E_{csurf} = E_c - q(\psi_{cnt,s} - \phi_0), \tag{3.27}$$

where E_{csurf} is the conduction band energy at the interface of the gate oxide and carbon nanotube. The Fermi level at the interface, E_{Fsurf} is shifted by an amount of V_{cb} and is expressed as [25,37]

$$E_{Fsurf} = E_F - qV_{cb} \qquad (3.28)$$

From Eqs. (3.27) and (3.28) we can write

$$E_{Fsurf} - E_{csurf} = E_F + q(\psi_{cnt,s} - V_{cb} - \phi_0) - E_c \qquad (3.29)$$

The shift in the Fermi level of the carbon nanotube due to impurity doping can be obtained as follows. The carrier concentration in a doped carbon nanotube is given by [40,41]

$$n_{cnt} = N_c I e^{\frac{E_F - E_c}{kT}} \qquad (3.30)$$

The intrinsic carrier concentration can be obtained by setting $E_F = 0$ in Eq. (3.30):

$$n_{cnt,i} = N_c I e^{\frac{-E_c}{kT}} \qquad (3.31)$$

Using Eq. (3.31) we can rewrite Eq. (3.30) as follows:

$$E_F = kT \ln\left(\frac{n_{cnt}}{n_{cnt,i}}\right) \qquad (3.32)$$

The carrier concentration in a doped carbon nanotube can also be obtained by adding an ionized impurity doping concentration, N in Eq. (3.31) as follows:

$$n_{cnt} = n_{cnt,i} + N. \qquad (3.33)$$

The ionized impurity concentration, N in Eq. (3.33) can be either donor atoms, N_D, or acceptor atoms, N_A.

In an intrinsic carbon nanotube, the Fermi level lies in the middle of the bandgap; we can use Eq. (3.32) to express the shift in the Fermi level depending upon the doping. Furthermore, using Eq. (3.33) we can define a parameter ΔE_F given by

$$\Delta E_F = \pm kT \ln\left(1 + \frac{N}{n_{cnt,i}}\right), \qquad (3.34)$$

where ΔE_F is positive for an n-type carbon nanotube (donors impurity concentration, $N = N_D$) and negative for a p-type carbon nanotube (acceptors impurity concentration ($N = N_A$). Equation (3.34) is valid for $N \le n_{cnt,i}(e^{Eg/2kT} - 1)$ and the impurity concentration, N is limited by the maximum value of $|\Delta E_F|$, which can bedetermined from the size of the energy gap as $|\Delta E_F| \le E_g/2$. As an example, for chiral vectors (11,3) and (5,3), using Eq. (3.34) and data in Table 2 of [41], the computed doping concentrations correspond to 5.438×10^{20} cm^{-3} and 7.299×10^{14} cm^{-3}, respectively. Thus, using Eq. (3.34) for the shift in the Fermi level, we can rewrite Eq. (3.29) as follows:

$$E_{Fsurf} - E_{csurf} = \Delta E_F + q(\psi_{cnt,s} - V_{cb} - \phi_0) - E_c \tag{3.35}$$

Using Eq. (3.35) in the analytical expressions for the charge inside the carbon nanotube given by Eqs. (3.17) and (3.18), the gate substrate voltage, V_{gb}, in Eq. (3.10) can be rewritten as

$$V_{gb} = \psi_{cnt,s} - \delta f(\psi_{cnt,s}, V_{cb}) + V_{fb}, \tag{3.36}$$

where

$$f(\psi_{cnt,s}, V_{cb}) = \begin{cases} Ie^{\frac{\Delta E_F + q(\psi_{cnt,s} - V_{cb} - \phi_0) - E_c}{kT}}; \\ \text{for } \psi_{cnt,s} \le V_{cb} + \phi_0 - \frac{\Delta E_F}{q} + \frac{E_c}{q} - \frac{kT}{q} \\ \frac{\sqrt{(\Delta E_F + q\psi_{cnt,s} - qV_{cb} - q\phi_0)^2 - E_c^2}}{kT}; \\ \text{for } \psi_{cnt,s} \ge V_{cb} + \phi_0 - \frac{\Delta E_F}{q} + \frac{E_c}{q} + \frac{kT}{q} \end{cases} \tag{3.37}$$

and

$$\delta = \frac{qLN_c}{C_{ox1}}. \tag{3.38}$$

Equation (3.36) cannot be solved explicitly in terms of the terminal voltages and numerical techniques are to be used to find the exact carbon nanotube potential given a gate input voltage.

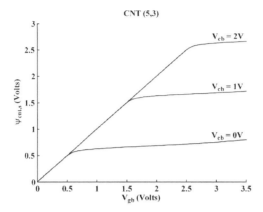

Figure 3.3 Carbon nanotube potential, $\psi_{cnt,s}$ versus gate substrate voltage, for $V_{fb} = 0$ and $\phi_0 = 0$ of a CNT-FET (5,3) using a numerical approach. The device dimensions are as follows: $T_{ox1} = 40$ nm, $T_{ox2} = 400$ nm and $L = 50$ nm.

Figure 3.3 shows the carbon nanotube potential versus the gate substrate voltage for varying V_{cb}. Nevertheless, by partitioning the Eq. (3.36) in three regions, an explicit solution in terms of the terminal voltages can be obtained: Region 1, in which the carbon nanotube potential has a linear and exponential dependence on the gate to substrate voltage, Region 3, in which the carbon nanotube potential does not change significantly with the gate to substrate voltage, and Region 2, in which no real dependence of the carbon nanotube potential on gate to substrate voltage can be established and an approximate curve fitting can be obtained. The carbon nanotube potential can be found using the following equations:
Region 1,
for

$$0 \leq V_{gb} \leq V_{cb} + V_{fb} + \phi_0 - \frac{\Delta E_F}{q} + \frac{E_c}{q} - \frac{kT}{q} - \frac{Ie^{-1}}{m}$$

$$\psi_{cnt,s} = V_{gb} - V_{fb},$$

$$(3.39)$$

Regions 2 and 3 (by using a linear polynomial fitting equation with slope m),
for

$$V_{gb} \geq V_{cb} + V_{fb} + \phi_0 - \frac{\Delta E_F}{q} + \frac{E_c}{q} - \frac{kT}{q} - \frac{Ie^{-1}}{m}$$

$$\psi_{cnt,s} = \frac{V_{gb} - \delta I e^{-1} - V_{fb} + \delta m \left(V_{cb} + \phi_0 - \frac{\Delta E_F}{q} + \frac{E_c}{q} - \frac{kT}{q} \right)}{1 + \delta m}, \quad (3.40)$$

where *m* is the slope of Region 2 and is of the following form:

$$m = \frac{\sqrt{\frac{2E_c}{kT} + 1} - Ie^{-1}}{\frac{2kT}{q}} \quad (3.41)$$

Figures 3.4a,b show a plot of $\psi_{cnt,s}$ versus V_{gb} for two chiral vectors (11,3) and (7,2), respectively. The solid lines in Fig. 3.4 correspond to the analytical solution and the numerical solutions are shown by circle markers. It is clearly noticed that the analytical solution agrees closely to the numerical solution. In Fig. 3.4,

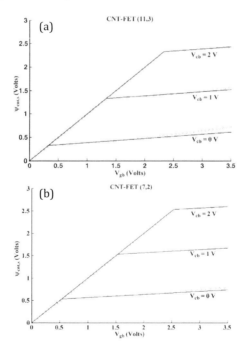

Figure 3.4 Carbon nanotube potential, $\psi_{cnt,s}$ versus gate substrate voltage using $V_{fb} = 0$ and $\phi_0 = 0$ of (a) CNT-FET (11,3) and (b) CNT-FET (7,2). The device dimensions are as follows: $T_{ox1} = 40$ nm, $T_{ox2} = 400$ nm and $L = 50$ nm.

the curve fitting has been used for the Region 2 given by the Eq. (3.40). The Region 2, defined by the gap predicted by the limits of Eqs. (3.17) and (3.18), is further extended to cover the Region 3. This is a very good approximation, but starts failing for gate voltages above 3 V. However, transistors will hardly operate beyond this voltage given the low power restrictions of the current CMOS technology [45–47].

3.2.1   Current Equation

In a CNT-FET, both diffusion and drift carrier transport mechanisms contribute to the current. We have considered both diffusion and drift transport mechanism since fabricated CNT-FETs [15,22,23] have carbon nanotube lengths longer than the magnitude of the optical phonon mean free path (100 nm) [48,49]. Thus, phonon scattering is present and ballistic transport cannot be considered in such devices. The non-equilibrium Green function (NEGF) approach is applied mostly to ballistic transport and can also be applied to current transport with some scattering present [51]. However, we have not considered NEGF approach for analytical simplicity. In CNT-FETs, CNT is formed by rolling a grapheme sheet into a tubular structure. Current is confined mainly to its circumference which can be described by the following equation of the form [37,50]

$$I_{ds} = I_{diff}(x) + I_{drift}(x) = \frac{|R|}{2L}\left[\int_{\psi'_{cnt}(0)}^{\psi'_{cnt}(L)} \mu(-Q'_{cnt})d\psi_{cnt} + \frac{kT}{q}\int_{Q'_{cnt}(0)}^{Q'_{cnt}(L)} \mu\, dQ'_{cnt}\right]. \quad (3.42)$$

$|R|$ in Eq. (3.42) is the circumference of the nanotube and μ is the carrier mobility in a carbon nanotube. In Eq. (3.42), we have used the charge per unit area, but only to show that we have considered a surface area, $|R|L/2$ for the carbon nanotube. The mobility can be replaced by $\mu_{eff} = \gamma\mu_{graphite}$, as each CNT (n,m) will have different values for the mobility, γ is a conversion factor for CNT from graphite with a value varying from 0 to 1. In addition, γ can also be used to represent how much surface area of the CNT is responsible for the charge flow. The charge, Q_{cnt} can be obtained from Eq. (3.13) as follows:

$$Q_{cnt} = C_{ox1}(-V_{gb} + \psi_{cnt,s} + V_{fb}). \quad (3.43)$$

Equation (3.43) can be expressed in terms of the charge per unit area by diving it by $|R|L/2$ to obtain Q'_{cnt}. By substituting Eq. (3.43) in Eq. (3.42), the following expression for the current is obtained:

$$I_{ds} = \beta[f\{\psi_{cnt,s}(L), V_{gs}\} - f\{\psi_{cnt,s}(0), V_{gs}\}], \qquad (3.44)$$

where

$$f\{\psi_{cnt,s}(x), V_{gs}\} = \left(V_{gs} + V_{sb} - V_{fb} + \frac{kT}{q}\right)\psi_{cnt,s}(x) - \frac{1}{2}\psi_{cnt,s}^2(x), \quad (3.45)$$

and

$$\beta = \gamma \frac{\mu\, C_{ox1}}{L^2}. \qquad (3.46)$$

In Eq. (3.44), $\psi_{cnt,s}(L)$ is the channel potential at the drain end, which can be found from Eqs. (3.39) and (3.40) by setting $V_{cb} = V_{ds} + V_{sb}$. The channel potential at the source end, $\psi_{cnt,s}(0)$ is obtained from Eq. (3.40) by setting $V_{cb} = V_{sb}$. In order for this assumption to be valid, the gate to substrate voltage must satisfy the condition $V_{gb} \geq V_{sb} + V_{fb} + \phi_0 - \frac{\Delta E_F}{q} + \frac{E_c}{q} - \frac{kT}{q}\frac{le^{-1}}{m}$. The right hand side of this inequality equation can be recognized as the threshold voltage. In addition, depending on which equation is to be used, either Eq. (3.39) or (3.40) to find the carbon nanotube potential, $\psi_{cnt,s}(L)$, two regions of operation can be defined as follows: a saturation region for $V_{ds} \geq V_{gs} - \left(V_{fb} + \phi_0 - \frac{\Delta E_F}{q} + \frac{E_c}{q} - \frac{kT}{q} - \frac{le^{-1}}{m}\right)$, and a linear region for $V_{ds} \leq V_{gs} - \left(V_{fb} + \phi_0 - \frac{\Delta E_F}{q} + \frac{E_c}{q} - \frac{kT}{q} - \frac{le^{-1}}{m}\right)$. The right hand side of this inequality equation is similar to a saturation voltage in a typical MOSFET and under parenthesis is the threshold voltage, V_{th} term. The saturation voltage, $V_{ds,sat}$ and V_{th} of a CNT-FET are described as follows:

$$V_{th} = V_{fb} + \phi_0 - \frac{\Delta E_F}{q} + \frac{E_c}{q} - \frac{kT}{q} - \frac{le^{-1}}{m}, \qquad (3.47)$$

$$V_{ds,sat} = V_{gs} - (V_{fb} + \phi_0 - \frac{\Delta E_F}{q} + \frac{E_c}{q} - \frac{kT}{q} - \frac{le^{-1}}{m}). \qquad (3.48)$$

Equation (3.44) can be easily modified to account for variation of current with V_{ds} in saturation region by introducing a parameter, λ which is equivalent to channel length modulation parameter in a MOSFET. The modified Eq. (3.44) is as follows:

$$I_{ds} = \beta [\, f\{\psi_{cnt,s}(L), V_{gs}\} - f\{\psi_{cnt,s}(0), V_{gs}\}](1 + \lambda V_{ds}) \qquad (3.49)$$

Figure 3.5 shows the *I–V* characteristics for a carbon nanotube with chiral vector (11,9) for different overdrive gate voltages obtained from Eq. (3.44) and (3.49). In Fig. 3.5, experimentally measured data taken from [15] for chiral vector (11,9) are also plotted for the comparison which follow very closely the modeled behavior. Figure 3.6 shows a plot of I_{ds} versus the gate to source voltage, V_{gs} of a CNT-FET with chiral vector (11,9) for varying V_{ds} values. The curves show that for a given V_{gs} the current I_{ds} increases with increasing V_{ds}, in saturation region as in Fig. 3.5.

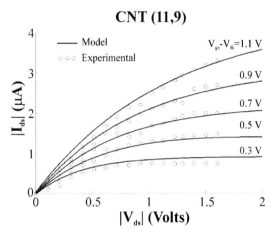

Figure 3.5   *I–V* characteristics of CNT-FET (11,9) with V_{fb} = -0.79 V and ϕ_0 = 0. The device dimensions are T_{ox1} = 15 nm, T_{ox2} = 120 nm and L = 250 nm. In modeled curve, $Q_{01} = Q_{02} = 0$ and λ = 0.1 V^{-1}.

3.3   Logic Gates Modeling

CNT-FETs can be made both n- and p-types as in CMOS [13], making possible the implementation of CNT-FETs as fully complementary logic, such as the inverters, NOR and NAND gates as shown in Fig. 3.7. The model equations characterizing the current voltage transport described in Section 3.2 are for n-type CNT-FETs and can also be used for p-type CNT-FETs by changing the polarities of the voltages as in a standard p-MOSFET. In generating voltage transfer characteristics, we have used Eqs. (3.39), (3.40), modified Eq. (3.44), i.e., Eq. (3.49), and two complementary CNT-FETs with two different chiral vectors.

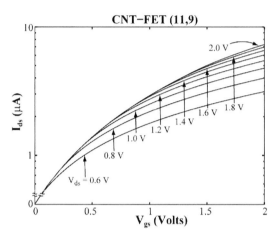

Figure 3.6   Current, I_{ds} versus gate to source voltage, V_{gs} using $V_{fb} = -0.79$ and $\phi_0 = 0$ for CNT-FET (11,9). The device dimensions are as follows: $T_{ox1} = 15$ nm, $T_{ox2} = 120$ nm and $L = 250$ nm.

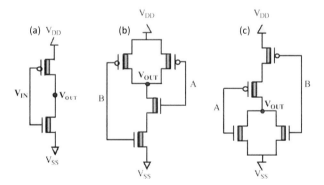

Figure 3.7   CNT-FET logic: (a) Inverter, (b) two input NAND gate and (c) two input NOR gate.

Inverter, NAND and NOR constitute basic building blocks in designing of digital integrated circuits. Understanding of their voltage transfer characteristics gives information about high and low logic levels, transition region, and noise margins. Using the current transport model we have modeled voltage transfer characteristics of basic gates using complementary CNT-FETs. Figure 3.8 shows the voltage transfer characteristic of an inverter for a chiral vector (11,9). The dotted line in Fig. 3.8 is the experimental curve plotted from the work of Derycke et al. [22] and Martel et al. [23] for comparison. The two solid lines in Fig. 3.8 correspond to λ = 0.1 and 0 V^{-1} with and without channel length modulation, respectively. It should be noted in Fig. 3.8 that V_{IN} varies from –2 to 2 V and V_{OUT} varies from –2 to 1.5 V, which correspond to conditions of the experiment. Therefore, modeled curves have been obtained for conditions of the experiment for comparison. It is seen from the Fig. 3.8 that the experimental voltage transfer characteristics closely follow the modeled voltage transfer characteristic corresponding to channel length modulation, λ = 0.1 V^{-1}. Experimental transfer characteristic shows some deviation from the modeled behavior which may be attributed to influence of process variation on device parameters.

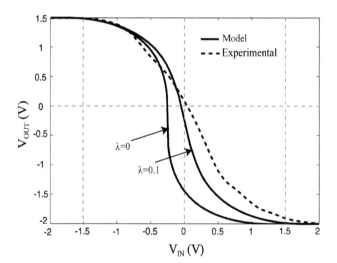

Figure 3.8 Voltage transfer characteristics of an inverter using CNT-FETs (11,9) with V_{fb} = 0 V, λ = 0, 0.1 V^{-1} and ϕ_{ms} = 0. The dimensions of both the n-type CNT-FET and p-type CNT-FET are as follows: T_{ox1} = 15 nm, T_{ox2} = 120 nm and L = 250 nm.

Figure 3.9 shows the voltage transfer characteristics of an inverter and two input NAND gate for a chiral vector (11,9). Figure 3.10 shows the voltage transfer characteristics of an inverter and a NOR gate for a chiral vector (11,9). The voltage transfer characteristic of the inverter is included to show its full output voltage. The power supply voltage is 2 V. The input voltage is varied from 0 to 2 V and corresponding output voltage is obtained. In NAND and NOR gates, voltage transfer characteristics have been obtained for the following input conditions: setting one of the inputs to high (V_{DD}) or low (V_{SS}) and varying the other input or varying both of the inputs simultaneously, as shown in Figs. 3.9 and 3.10. It is seen that the inverter, NOR and NAND gates give full logic swing similar to inverter and gates designed in CMOS. The inverter switching threshold voltage is 1 V. In NAND and NOR gates, the switching threshold voltage is dependent on input voltage conditions. The voltage transfer characteristics in these gates also exhibit sharp transition at switching thresholds and are similar to characteristics observed in corresponding gates implemental in CMOS.

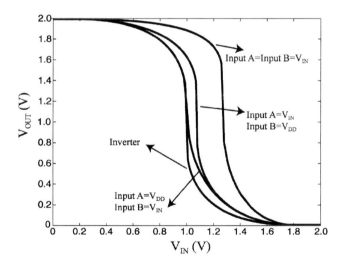

Figure 3.9   Voltage transfer characteristics of an inverter and a NAND gate using CNT-FETs (11,9) with $V_{fb} = 0$ V and $\phi_0 = 0$. The dimensions of both the n-type CNT-FET and p-type CNT-FET are as follows: $T_{ox1} = 15$ nm, $T_{ox2} = 120$ nm and $L = 250$ nm.

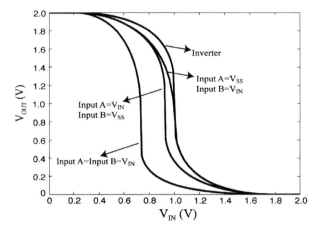

Figure 3.10   Voltage transfer characteristics of an inverter and a NOR gate using CNT-FETs (11,9) with $V_{fb} = 0$ V and $\phi_0 = 0$. The dimensions of both the n-type CNT-FET and p-type CNT-FET are as follows: $T_{ox1} = 15$ nm, $T_{ox2} = 120$ nm and $L = 250$ nm.

3.4   Conclusion

In the present work, the charge transport in a CNT-FET, based on our recently reported carrier concentration model, has been used to relate the carbon nanotube potential and the gate substrate voltage. Analytical solutions have been developed relating the carbon nanotube potential and the gate substrate voltage. These solutions are then used to model the current transport in a CNT-FET depending on the chiral vector and device geometry analytically. Threshold and saturation voltage model equations are each derived in the process. The analytical transport models have been used to generate CNT-FET *I–V* characteristics and compared with the recently reported experimental data for the chiral vector (11,9). A close agreement is obtained between the analytical models and the experimental observations in linear and saturation regions.

The analytical model equations for the current transport have been used to characterize voltage transfer characteristics of complementary logic devices such as the inverter, NAND and NOR gates. The voltage transfer characteristic of the CNT-FET inverter is similar to the voltage transfer characteristic of a typical CMOS inverter and shows a sharp transition at the inverter logic threshold voltage, 1.0 for a 2 V operation. Transfer characteristics are also compared with the experimentally measured transfer characteristics for a chiral vector (11,9). The gates, NAND and NOR are also simulated and show a sharp transition in their voltage transfer characteristics for a 2 V operation, which are similar to voltage transfer characteristics of typical CMOS gates.

The analytical transport model equation for the CNT-FET can be expressed in terms of SPICE compatible model parameters. Furthermore, using the charge and potential modeling used in this work, a more accurate current equation for shorter carbon nanotubes, for instance using ballistic transport behavior and NEGF approach can be employed and is suggested for future work.

References

1. Iijima, S. (1991). Helical microtubules of graphitic carbon, *Nature*, **354**, 56–58.
2. Cui, Y, Wei, Q., Park, H., and Lieber, C. (2001). Nanowire nanosensors for highly-sensitive, selective and integrated detection of biological and chemical sensors, *Science*, **293**, 1289–1292.
3. Baughman, R. H., Zakhidov, A. A., and de Heer, W. A. (2002). Carbon nanotubes—the route towards applications, *Science*, **297**, 787–792.
4. Srivastava, A., Soundararajan, R., and Hsu, J.-C. (2006). CMOS chip chemical detection system comprising mass-sensitive nanocantilevers, **6172**, 617200Y-1 to 61720Y-10.
5. Tans, S. J., Vershueren, A. R. M., and Dekker, C. (1998). Room-temperature transistor based on a single carbon nanotube, *Nature*, 393, 49–52.
6. Martel, R., Schmidt, T., Shea, H. R., Hertel, T., and Avouris, P. (1998). Single- and multi-wall carbon nanotube field-effect transistors, *Appl. Phys. Lett.*, **73**, 2447–2450.
7. Avouris, P., Appenzeller, J., Martel, R., and Wind, S. J. (2003). Carbon nanotube electronics, *Proc. IEEE*, **91**, 1772–1784.
8. DeHon, A., and Likharev, K. K. (2005). Hybrid CMOS/nanoelectronic digital circuits: Devices, architectures, and design automation, *Proceedings of International Conference on Computer-Aided Design* (*ICCAD'05*), pp. 375–382.
9. Srivastava, N., Joshi, R. V., and Banerjee, K. (2005). Implications for performance, power dissipation, and thermal management, *IEDM Tech. Digest*, 249–252.

10. Raychowdhury, A., and Roy, K. (2006). Modeling of metallic carbon-nanotube interconnects for circuit simulations and a comparison with Cu interconnects for scaled technologies, *IEEE Trans. Comput.-Aided Des. Integrated Circuits Syst.*, **25**, 58–65.

11. Dresselhaus, M. S., Dresselhaus, G., and Avouris, P. (2001). *Carbon Nanotube: Synthesis, Properties, Structure, and Applications* (Springer Verlag).

12. Wildoer, J., Venema, L., Rinzler, A., Smalley, R., and Dekker, C. (1998). Electronic structure of atomically resolved carbon nanotubes, *Nature*, **391**, 59–62.

13. Wong, H. S. P. (2002). Field-effect transistors—from silicon MOSFET to carbon nanotube FETs, *Proceedings of 23th International Conference on Microelectronics* (*MIEL*), pp. 103–107.

14. Guo, J., Datta, S., Lundstrom, M., Brink, M., McEuen, P., Javey, A., Dai, H., Kim, H., and McIntyre, P. (2002). *IEDM Technical Digest*, 711–714.

15. Wind, S. J., Appenzeller, J., Martel, R., Derycke, V., and Avouris, P. (2002). Vertical scaling of carbon nanotube field-effect transistors using top gate electrodes, *Appl. Phys. Letts.*, **80**, 3817–3819.

16. Javey, A., Guo, J., Farmer, D. B., Wang, Q., Wang, D., Gordon, R. G., Lundstrom, M., and Dai, H. (2004). *Nano Lett.*, **4**, 447–450.

17. Wind, S. J., Appenzeller, J., Martel, R., Derycke, V., and Avouris, P. (2002). Fabrication and electrical characterization of top gate single-wall CNFETs (2002). *J. Vac. Sci. Technol. B*, **20**, 2798–2801.

18. Nosho, Y., Ohno, Y., Kishimoto, S., and Mizutani, T. (2005). N-type carbon nanotube field-effect transistors fabricated by using Ca contact electrodes, *Appl. Phys. Letts.*, **86**, 073105–073107.

19. Javey, A., Wang, Q., Kim, W., and Dai, H. (2003). Advances in complementary carbon nanotube field-effect transistors, *IEDM Tech. Dig.*, 31.2.1–31.2.4.

20. Javey, A., Wang, Q., Ural, A., Li, Y., and Dai, H. (2002). Carbon nanotube transistors arrays for multistage complementary logic and ring oscillators, *Nano Letts.*, **2**, 929–932.

21. Javey, A., Kim, H., Brink, M., Wang, Q., Ural, A., Guo, J., Mcientyre, P., McEuen, P., Lundstrom, M., and Dai, H. (2002). High k-dielectric for advanced carbon nanotube transistors and logic gates, *Nat. Mater.*, **1**, 241–246.

22. Derycke, V., Martel, R., Appenzeller, J., and Avouris, P. (2001). Carbon nanotube inter- and intramolecular logic gates, *Nano Letts.*, **1**, 453–456.

23. Martel, R., Derycke, V., Appenzeller, J., Wind, S., and Avouris, P. (2002). Carbon nanotube field effect transistors and logic circuits, *Proceedings of 39th Design Automation Conference*, 2002, pp. 94–98.

24. Bachtold, A., Hadley, P., Nakanishi, T., and Dekker, C. (2001). Logic circuits with carbon nanotube transistors, *Science*, **294**, 1317–1320.

25. John, D. L., Castro, L. C., Clifford, J. P., and Pulfrey, D. L. (2003). *IEEE Trans. Nanotechnol.*, **2**, 175–180.

26. Dwyer, C., Cheung, M., and Sorin, D. J. (2004). Semi-empirical SPICE models for carbon nanotube FET logic, *Proceedings of 4th IEEE Conference on Nanotechnology*, 386–388.

27. Raychowdhury, A., Mukhopadhyay, S., and Roy, K. (2004). A circuit-compatible model of ballistic carbon nanotube field effect transistors, *IEEE Trans. Comput.-Aided Des. Integrated Circuits Syst.*, **23**, 1411–1420.

28. Raychowdhury, A., and Roy, K. (2005). Carbon nanotube based voltage-mode multiple-valued logic design, *IEEE Trans. Nanotechnol.*, **4**, 168–179.

29. Connor, I. O', Liu, J., Gaffiot, F., Prégaldiny, F., Lallement, C., Maneux, C., Goguet, J., Frégonèse, S., Zimmer, T., Anghel, L., Dang, T.-T., and Leveugle, R. (2007). *IEEE Trans. Circuits Syst. Part*-1, **54**, 2365–2379.

30. Hazeghi, A., Krishnamohan, T., and Wong, H.-S. P. (2007). Schottky-barrier carbon nanotube field effect transistor modeling, *IEEE Trans. Electron Devices*, **54**, 439–445.

31. Burke, P. J. (2004). AC performance of nanoelectronics: Towards a ballistic THz nanotube transistor, *Solid-State Electron.*, **48**, 1981–1986.

32. Burke, P. J. (2003). An RF circuit model for carbon nanotubes, *IEEE Trans. Nanotechnology*, **2**, 55–58.

33. Castro, L. C., and Pulfrey, D. L. (2006). Extrapolated f_{max} for carbon nanotube field-effect transistors, *Nanotechnology*, **17**, 300–304.

34. Castro, L. C., Pulfrey, D. L., and John, D. L. (2007). High-frequency capability of Schottky-barrier carbon nanotube FETs, *Solid-State Phenomena*, **121–123**, 693–696.

35. Srivastava, A., Marulanda, J. M., Xu, Y., and Sharma, A. K. (2009). Current transport modeling of carbon nanotube field effect transistors, *Phys. Status Solidi (A)*, **206**, 1569–1578.

36. Shen, L. C., and Kong, J. A. (1995). *Applied Electromagnetism* (PWS Foundations in Engineering Series, Boston, MA).

37. Tsividis, Y. (1999). Operation and Modeling of the MOS transistor (McGraw-Hill, Singapore).

38. Hayt, W. H. (1974). *Engineering Electro-Magnetics* (McGraw-Hill, New York).

39. Marulanda, J. M., Srivastava, A., and Nahar, R. K. (2005). *Proceedings of 13th International Workshop on the Physics of Semiconductor Devices* (*IWPSD*), New Delhi, pp. G-11.

40. Marulanda, J. M., and Srivastava, A. (2007). Carrier density and effective mass calculations for carbon nanotubes, *Proceedings of International Conference on Integrated Circuit Design & Technology* (*ICICDT*), Austin, TX, pp. 234–237.

41. Marulanda, J. M., and Srivastava, A. (2008). Carrier density and effective mass calculations in carbon nanotubes, *Phys. Status Solidi* (*B*), **245**, 2558–2562.

42. Saito, R., Dresselhaus, M. S., and Dresselhaus, G. (1998). *Physical Properties of Carbon Nanotubes* (Imperial College Press, London, U.K.)

43. Fregonese, S., Cazin d'Honincthun, H., Goguet, J., Maneux, C., Zimmer, T., Bourgoin, J. P., Dollfus, P., and Galdin-Retailleau, S. (2008). Computationally efficient physics-based compact CNT-FET model for circuit design, *IEEE Transactions on Electron Devices*, **55**, 1317–1327.

44. Paul, B. C., Fujita, S., Okajima, M., and Lee, T. (2006). Modeling and analysis of circuit performance of ballistic CNTFETS, *ACM/IEEE Proceedings Design Automation Conference*, pp. 717–722.

45. Allen, P. E., and Holberg, D. R. (2002). *CMOS Analog Circuit Design* (Oxford University Press, New York).

46. Chandrakasan, A., and Brodersen, R. W. (1998). *Low Power CMOS Design* (John Wiley & Sons Inc., New York).

47. Roy, K., and Prasad, S. (2000). *Low Power CMOS VLSI Circuit Design* (John Wiley & Sons Inc., New York).

48. Yao, Z., Kane, C. L., and Dekker, C. (2000). High-field electrical transport in single-wall carbon nanotubes, *Phys. Rev. Lett.*, **84**, 2941–2944.

49. Javey, A., Guo, J., Wang, Q., Lundstron, M., and Dai, H. (2003). Ballistic carbon nanotube transistors, *Nature*, **424**, 654–657.

50. Streetman, B. G. (2000). *Solid State Electronic Devices*, 5th ed. (Prentice Hall, India).

51. Datta, S. (2002). The non-equilibrium Green's function (NEGF) formalism: An elementary introduction, *IEDM Tech. Dig.*, 703–706.

Chapter 4

Single-Walled Carbon Nanotube Interconnection

4.1 Introduction

The one-dimensional carbon nanotube (CNT) has excellent electrical, mechanical, and thermal properties [1,2], which has made the CNT one of the promising materials for applications in nanoelectronics [3–6] and micro/nano-systems [7]. In nanoelectronics, the carbon nanotube field-effect transistor (CNT-FET) is very promising in design of emerging logic devices for nanoscale integration and there is a noticeable amount of published and ongoing research on understanding current transport in CNT-FETs and developing models for use in circuit simulators [8–17]. In micro/nano-systems, both the CNT and CNT-FET are very promising as sensors for detecting chemicals, gases at molecular levels [7,18–22]. Carbon nanotube carries a current density of $\sim 10^{10}$ A/cm^2, which is higher by a two to three orders of magnitude in Cu. Its mean free path is in micrometer range compared to ~ 40 nm mean free path in Cu. The large mean fee path in the CNT allows a ballistic transport over a wider range of micrometers resulting in reduced resistivity, and strong atomic bonds [23] provide tolerance to electromigration [24,25]. Higher thermal conductivity makes the CNT suitable for use in tall vias of 3D ICs [26–28]. Electrical performance of single and bundled carbon

Carbon-Based Electronics: Transistors and Interconnects at the Nanoscale
Ashok Srivastava, Jose Mauricio Marulanda, Yao Xu, and Ashwani K. Sharma
Copyright © 2015 Pan Stanford Publishing Pte. Ltd.
ISBN 978-981-4613-10-1 (Hardcover), 978-981-4613-11-8 (eBook)
www.panstanford.com

nanotubes have been studied in the work of Plombon et al. [29], Yao et al. [30] and Nougaret et al. [31]. Recently, Sarto and Tamburrano [32] have presented analytical derivation of multi-walled carbon nanotubes (MWCNTs) from the multiconductor transmission line model. Properties of carbon nanomaterials relevant to VLSI interconnects which include single-walled carbon nanotube (SWCNT), MWCNT, graphene nanoribbons (GNRs), and comparison with the properties of Cu interconnect are summarized in the work of Li et al. [26]. Graphene nanoribbons are a recent addition to the interconnect technology since the discovery of 2D graphene in 2004 [33,34] and the methods of fabricating GNR are still being developed.

Nanometer CMOS technology, especially in 22 nm and below, is plagued due to performance degradation of conventional Cu/low-k dielectric as an interconnect material for gigascale integration. In one of the recent published research studies on interconnect technologies, Koo et al. [24] have mentioned the effect of scaling on surface and grain boundary scattering and electromigration in Cu interconnect [35] and in great detail degradation in its parameters such as the latency and power dissipation. Thus, the need for other materials possibly substituting Cu/low-k dielectric interconnections has brought forward other novel interconnect technologies for next-generation VLSI interconnects. Optical interconnects have already been suggested for on-chip integration [35–38] but still face serious integration problems. Among newer and novel VLSI interconnection technologies, CNTs and GNRs have emerged as promising candidates for next-generation VLSI interconnects [4–6,25,39–42]. An excellent review of these technologies has been presented in one of the recent publications of Li et al. [26]. Although the optical interconnect is still being investigated due to its inherent advantages over the Cu interconnect, other new technologies such as the capacitively driven low-swing interconnect (CDLSI) have been also evolved [24]. In search for novel technologies, no such material has aroused so much interest other than carbon nanomaterials since the discovery of carbon nanotube in 1991 by Iijima [43].

A model describing the electromagnetic field propagation along a CNT is indispensable in order to study the interconnection performance of CNT while comparing with traditional metal interconnects. Three theories are used to build different models.

Lüttinger liquid theory [44] describes interacting electrons (or other fermions) in a one-dimensional conductor and is necessary since the commonly used Fermi liquid model breaks down in one-dimension. Burke [45,46] regards that electrons are strongly correlated when they transport along the CNT and proposed a transmission line model based on the Lüttinger liquid theory. Another transmission line model was built based on the Boltzmann transport equation (BTE) [47]. Two-dimensional electron gas, where the charged particles are confined to a plane and neutralized by an inert uniform rigid positive plane background was studied by Fetter [48,49]. Based on the work of Fetter [48,49], Maffucci et al. [50] investigated electron transport along the CNT and proposed a third model, fluid model. In these models [48–50], electron–electron correlation, which is significant in CNTs [51–53], has not been considered. The first model is based on quantum dynamics concepts; the second model requires solving the BTE; the third model has been developed within the framework of the classical electrodynamics and is simple on concepts and mathematical modeling.

Interacting electrons in two and three-dimensions are well described in terms of an approximate model of weakly interacting quasi-particles, namely Fermi liquid theory. This model has been highly successful in explaining the properties of two and three-dimensional conductors. However, this approximate picture does not hold in one-dimension. Instead, the ground state of an interacting one-dimensional electron gas (1DEG) is a strongly correlated state known as a Lüttinger liquid. Unlike in a Fermi liquid, in a Lüttinger liquid, the low energy excitations are Bosonic sound-like density waves (plasmas). In two-dimensional electron gas model [48–50], electron–electron correlation has not been considered in studying electron transport in CNTs. Although the Lüttinger liquid model using quantum mechanical concept considers this correlation, the result and expression are too complicated to be solved. As quasi one-dimensional system, quantum effects [54] must be considered to characterize CNT interconnects. Therefore, we have made modification to the two-dimensional electron gas model to include electron–electron interactions and built one-dimensional liquid model [55], which is relatively easy to solve and apply in transmission line SWCNT interconnect modeling. The model can be easily used to study

S parameters and group delays of SWCNT interconnect for RF/ microwave applications. In the following sub-sections, we will first describe the two-dimensional fluid model and our one-dimensional fluid model of SWCNTs.

4.2   Two-Dimensional Fluid Model

An SWCNT is one-atom-thick sheet of graphite (called graphene) rolled up into a seamless cylinder with diameter of the order of a nanometer. This results in a nanostructure where the length-to-diameter ratio exceeds 10,000. Since carbon nanotubes are constructed of hexagonal networks, the carbon atoms contain sp^2 hybridization. There are four valence electrons for each carbon atom. The first three electrons belong to the σ orbital and are at energies 2.5 eV below the Fermi level; therefore, they do not contribute to the conduction. The fourth valence electron, however, is located in the π orbital, which is slightly below the Fermi level; therefore, this electron is very likely to control conduction and transport properties. This electron corresponds to the valence band of the energy band diagram. The anti-bonding π orbital is slightly above the Fermi level, which corresponds to the conduction band in an energy band diagram. Depending upon the direction of the graphene sheet is wrapped; the SWCNT can be metallic or semiconducting.

If the graphene sheet rolled to form the SWCNT is infinitesimally thin, then the conduction electrons are distributed on the lateral surface, s' of the SWCNT cylinder shell and the electrons are embedded in a rigid uniform positive charge background with a uniform surface number density. Thus, the motion of the electrons is confined to the surface. Furthermore, electrical charge neutrality requires that in equilibrium the conduction electron charge density precisely cancels with that of the background positive ions. According to this analysis, two-dimensional fluid model could be utilized to study the electron transport along the SWCNT. This model is shown in Fig. 4.1. The cylinder shell radius is r and length is l. The cylinder axis is oriented along the z-axis of the reference system. Two assumptions have been made to utilize this model. One is the electrons can only move along the z-axis; other is that all other fluid variables, such as the tangential component of the electric field to the nanotube surface, s', are almost uniform in the cross section of the SWCNT. These two assumptions are valid if both the nanotube length and the smallest wavelength of the electromagnetic field are much greater than the nanotube radius [56,57].

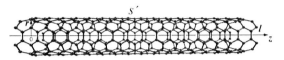

Figure 4.1 Geometry of a single-walled carbon nanotube (SWCNT).

The basic equation in fluid model is Euler's equation with Lorentz force term, which is Newton's second law applied in fluid dynamics [48] and is given by

$$mN\left(\frac{\partial}{\partial t} + \vec{V} \cdot V\right)\vec{V} = -\nabla\vec{P} - eN\vec{\mathcal{E}} - mN\nu\vec{V}, \qquad (4.1)$$

where $N(\vec{R}, t)$ is the electron three-dimension carrier density, $\vec{V}(\vec{R}, t)$ is the electron mean velocity, $\vec{R}$ is the position vector, $\vec{P}$ is the pressure, m is the electron mass, e is the electronic charge, and $\vec{\mathcal{E}}$ is electric field. The last term on the right-hand side represents the effect of scattering of electrons with the positive charge background and ν is the electron relaxation frequency. The relaxation frequency, ν is related to the mean-free path, λ, as follows:

$$\nu = \frac{v_F}{\lambda}, \qquad (4.2)$$

where v_F is the Fermi velocity. The value of λ is decided by the electron scattering mechanism. Experimental and theoretical studies have shown that there are two major scattering in a CNT [30]. One is the elastic scattering with acoustic phonons, which is only dependent on the material and is characterized by a constant mean-free path λ_e. Other is scattering with optical phonons, which is significant at high electric field intensity. It is characterized by the optical phonons mean-free path, λ_o and is given by

$$\lambda_o = \frac{h\nu_o}{eE}. \qquad (4.3)$$

In Eq. (4.3), ν_o is the frequency of phonons. Thus, the total mean free path, λ, is

$$\lambda^{-1} = \lambda_{e}^{-1} + \lambda_{o}^{-1}. \tag{4.4}$$

In fluid model, Eq. (4.1) is simplified to the following form:

$$mn\left(\frac{\partial}{\partial t} + v_z \frac{\partial}{\partial z}\right)v_z = -\frac{\partial p}{\partial z} - en\mathcal{E}_z\Big|_{s'} - mnv\,v_z, \tag{4.5}$$

where n is the electron density in this one-dimensional system, v_z is the electron mean velocity in z direction, p is the pressure in one-dimensional system. $\mathcal{E}_z$ is electric field in z direction along the SWCNT surface, s'. In Eq. (4.5), the variables have been changed to the lower case, which means that they are functions of the frequency and distributed on the SWCNT surface, s'.

4.3 One-Dimensional Fluid Model

Equations (4.1) and (4.5) describe the two-dimensional fluid and need to be modified to be used for carbon nanotubes, which are quasi one-dimensional systems. In a two-dimensional electron fluid, total energy is equal to the kinetic energy. In graphene sheet, the external electrical field drifts electrons in z direction while the electrons can also distribute in perpendicular (y) direction to form a two-dimensional fluid. This means that the whole energy provided by the external electrical field equals the two-dimensional electron fluid kinetic energy. However, when the graphene sheet is rolled to form a carbon nanotube, which is quasi one-dimensional system, the y direction shrinks into one point and the distributed electrons in y direction in a two-dimensional system will be in the same point in one-dimensional system. As a result, the repulsive force among the electrons will be significant. One-dimensional electron fluid energy can thus be considered to consist of two parts, the potential energy and the kinetic energy. The external electric field provides both the potential and kinetic energy to the one-dimensional fluid. Now we define a parameter α as follows:

$$\alpha \equiv \frac{E_{zP}}{E_z} = \frac{E_P}{E} = \frac{E_P}{E_K + E_P}, \tag{4.6}$$

where $\mathcal{E}_{zP}$ is the part of the electrical field that provides potential energy to electrons in z direction. E is the total energy of electrons. E_P and E_K are the potential and kinetic energies of electrons, respectively.

There are two channels in an SWCNT and two different spin electrons in each channel. We can consider that there are four electrons at the same point in one-dimensional SWCNT. We can then calculate the potential energy by moving these four electrons from ∞ to the same point of the SWCNT within ~1 nm diameter. The potential energy can be defined as follows [58]:

$$E_P = \sum_{n=2}^{4}(n-1)\times\frac{e^2}{2\pi\varepsilon_0}\frac{1}{d}\approx 18\,\text{eV}, \tag{4.7}$$

where d is the diameter of the SWCNT.

We assume that the velocity of these four electrons equals to the Fermi velocity. As a result, the kinetic energy is given by

$$E_K = 4\times\frac{1}{2}mv_F^2 \approx 7\,\text{eV}, \tag{4.8}$$

where $v_F = 3\gamma_0 b/2\hbar = 8\times10^5$ m/s is the Fermi velocity, γ_0 is the characteristic energy of the graphene lattice and is equal to 2.7 eV, and $\hbar$ is Planck constant. The inter-atomic distance, b in an SWCNT is 0.142 nm [50]. Substituting Eqs. (4.7) and (4.8) into Eq. (4.6), we obtain $\alpha \approx 0.7$.

It should be noted that the Lorentz force term, which belongs to body force terms in fluid dynamics, is a source momentum [59]. The external electric field provides both the potential and kinetic energy to the fluid. As a result, one-dimensional fluid model can be expressed as follows [55]:

$$mn\left(\frac{\partial}{\partial t}+v_z\frac{\partial}{\partial z}\right)v_z = -\frac{\partial p}{\partial z}-en\{(1-\alpha)\mathcal{E}_z|_s\}-mnvv_z, \tag{4.9}$$

The fluid model described by Eq. (4.9) assumes flow of one-dimensional electron fluid under the low external electric fields, $\mathcal{E}_z$. Yao, et al. [30] and Park et al. [60] have studied electronic transport in SWCNTs using low and high-resistance contacts under small and large bias-voltages and attempted to explain the conductance behavior from electron–phonon scattering. Yao et al. [30] have also observed linear *I–V* characteristics from measurement on samples using low-resistance contacts with slight deviation near 5 V from the linear behavior in some samples. Since current VLSI circuits use low-voltage nanometer CMOS technologies where bias voltage is below 1.5 V, deviation in the *I–V* relationship of CNT interconnect toward saturation at large bias-voltages should not be the cause of concern and so the applicability of Eq. (4.9) for CNT interconnects using low-resistance contacts.

The difference between our one-dimensional fluid model described in Eq. (4.9) with the two-dimensional fluid model [48–50] is the Lorentz force term. In a two-dimensional electron fluid [48], total energy is equal to the kinetic energy. In a graphene sheet, the external electrical field drifts electrons in z direction while electrons can also distribute in perpendicular (y) direction to form a two-dimensional fluid. This means that the whole energy provided by the external electrical field equals the two-dimensional electron fluid kinetic energy. However, when the graphene sheet is rolled to form a carbon nanotube, which is a quasi one-dimensional system, y direction shrinks into one point and the distributed electrons in y direction in a two-dimensional system will be at the same point in one-dimensional system. As a result, the potential energy due to repulsive force among the electrons will be significant. The external electric field provides both the potential and kinetic energies to the one-dimensional fluid.

Equation (4.9) describes electron transport in SWCNT and can be solved with the continuity equation on the surface, s':

$$\frac{\partial \sigma}{\partial t} + \frac{\partial j}{\partial z} = \frac{e \partial n}{\partial t} + \frac{\partial (e n v_z)}{\partial z} = 0, \tag{4.10}$$

where σ is the charge density and n is the electron density. The current density, $j = e n v_z$, and v_z is the electron velocity in the z direction.

The following equation, which assumes pressure, p, as the only function of the carrier density of conduction electrons, can be expressed as

$$p = p(n). \tag{4.11}$$

The metallic SWCNT is a good conductor, the electrical field on the surface will be nearly equal to zero. Using the perturbation method described in [48,50] that the system is initially in its equilibrium state and a perturbation is applied, we can write

$$n(z,t) = n_0 + \delta n, \qquad (4.12)$$

$$P(z,t) = p_0 + \delta P, \qquad (4.13)$$

where n_0 and p_0 are effective the density of conduction electrons and the pressure in thermal equilibrium, respectively. δn and δp are excess carrier concentrations of electrons and the excess pressure, respectively. Considering Eq. (4.11), we can write

$$\delta p = \delta n \frac{dp}{dn}\bigg|_{n=n_0} = m u_e^2 \delta n. \qquad (4.14)$$

In Eq. (4.14), $dp/dn|_{n=n_0} = m u_e^2$ is an abbreviate expression of the thermodynamic derivation [48] and u_e is the thermodynamic speed of sound of the electron fluid under neutral environment, i.e., under no charge present. The term, thermodynamic speed of sound is a thermodynamics term, which can be used to describe response of particles in fluid under external perturbation.

For two-dimensional fluid, total energy is identical to the kinetic energy [48] and is given by

$$E = E_K = \frac{1}{2}\left(\frac{n_0 \hbar^2 k^2}{2m}\right), \qquad (4.15)$$

where multiplier 1/2 is for the two-dimensional ideal spin −1/2 Fermi fluid. When we consider one-dimensional fluid, the total energy, E, is the sum of potential energy and kinetic energy. Combining Eqs. (4.6) and (4.15), we obtain

$$E = E_K + E_P = \frac{1}{1-\alpha}\left(\frac{n_0 \hbar^2 k^2}{4m}\right), \qquad (4.16)$$

where $k = (2\pi n_0)^{0.5}$ is the Fermi wave number [48]. The pressure is derivative of the total energy with respect to length in one-dimension and is obtained by substituting $k = (2\pi n_0)^{0.5}$ in Eq. (4.16):

$$p = \frac{1}{1-\alpha}\left(\frac{\hbar^2 \pi n_0^2}{2m}\right). \tag{4.17}$$

From Eqs. (4.11) and (4.17) and $k = (2\pi n_0)^{0.5}$, we obtain thermodynamic speed of sound, u_e as follows:

$$u_e = \frac{v_F}{\sqrt{2(1-\alpha)}}, \tag{4.18}$$

where $v_F = \hbar k/m_e$ is Fermi velocity [48]. It should be noticed that Eq. (4.18) is modified from classical $u_e = v_F/\sqrt{2}$ [48] and takes into account electron-election interactions. However, the fluid fails to reach the thermodynamic equilibrium in one oscillation period at higher frequencies. It is corrected by replacing 2 in denominator of Eq. (4.18) by $3/(D + 2)$ [48,56], where D is the number of dimensions. Here $D = 1$ for one dimension and Eq. (4.18) reduces to the following form:

$$u_e = \frac{v_F}{\sqrt{1-\alpha}} \tag{4.19}$$

In comparison to the mean-free path of electrons in Cu (40 nm) at room temperature, SWCNTs, on the other hand, have electron mean-free paths of the order of a micron [61]. Since the mean-free path of electrons in an SWCNT is close to 1 µm or less, the last term in Eq. (4.9) approaches zero.

Equation (4.9) can be further simplified. We assume that the electron fluid is initially in its equilibrium state and a perturbation is applied. Therefore, we consider first order of variables v_z and $\mathcal{E}_z$ and neglect higher orders. Substituting Eqs. (4.12), (4.13), and (4.14) into Eq. (4.9) and considering that the derivative of v_z with respected to z is zero, we obtain

$$mn_0\frac{\partial v_z}{\partial t} = -\frac{\partial \delta p}{\partial z} - en_0\{(1-\alpha)\mathcal{E}_z\} - mn\{\text{sgn}(l)\}vv_z, \tag{4.20}$$

where l is the length of SWCNT and $\text{sgn}(l)$ is the sign function defined as follows:

$$
\text{sgn}(l) = \begin{cases} 0 & \text{if} & l < l_{\text{mfp}} \\ 1 & \text{if} & l \geq l_{\text{mfp}} \end{cases},
$$

where l_{mfp} is the electron mean-free path in SWCNTs.

The parameter α in Eq. (4.20) describes the classical electron–electron repulsive force as described previously. In Section 4.5, we will see that the repulsive force has the same effects as the electron–electron correlation described in the Lüttinger liquid model [45,46], which is a quantum phenomenon in one-dimensional system.

The distributions of the surface charge density, σ, and induced current linear density, j, are described as follows [58]:

$$
\sigma(z,t) = -en(z,t) = -e\{n_0 - \delta n(z,t)\}, \tag{4.21}
$$

$$
j(z,t) = -env(z,t) = -e\{n_0 + \delta n(z,t)\}v(z,t), \tag{4.22}
$$

By combining Eqs. (4.14–4.22) and considering perturbation, δn, we obtain

$$
\frac{\partial j(z,t)}{\partial t} + vj + u_e^2 \frac{\partial \sigma(z,t)}{\partial z} = \frac{e^2 n_0}{m}(1-\alpha)\mathcal{E}_z. \tag{4.23}
$$

Equation (4.23) is the transport equation for one-dimension electron fluid in a metallic SWCNT. This equation together with Eq. (4.10) gives the relation of electric field, $\mathcal{E}_z$ with electric charge, σ, and current density, j.

There are several parameters in Eq. (4.23) that are still unknown. If the third term on the left-hand side of Eq. (4.23) is neglected since metallic SWCNT is a good conductor, we can obtain an equation for current density in frequency domain:

$$
\hat{j} = \frac{e^2 n_0}{m}(1-\alpha)\frac{1}{v + i\omega}\hat{\mathcal{E}}, \tag{4.24}
$$

where $\hat{j}$ and $\hat{\mathcal{E}}$ are, respectively, current density and electric field in frequency domain. Since $\hat{j} = \sigma\hat{\mathcal{E}}$, we obtain,

$$\sigma(\omega) = \frac{e^2 n_0}{m}(1-\alpha)\frac{1}{v+i\omega}.$$

(4.25)

In Eq. (4.25), σ is the frequency-dependent or dynamic conductivity. In microwave and infrared frequency regimes, the axial conductivity of metallic SWCNT is given by [62]

$$\tilde{\sigma}_{zz} = \frac{2e^2 v_F}{\pi^2 \hbar r}\frac{1}{v+i\omega},$$

(4.26)

where $\tilde{\sigma}_{zz}$ is the semi-classical version of the axial conductivity.

Equating Eqs. (4.25) and (4.26), we obtain

$$\frac{n_0}{m} = \frac{2v_F}{\pi^2 \hbar r}\left(\frac{1}{1-\alpha}\right).$$

(4.27)

Equation (4.27) is very useful in deriving parameters of SWCNT transmission line model, which is described in the following section.

4.4 Transmission Line Model

We consider a metallic SWCNT above a perfect conducting plane and assume [50] that the propagating EM wave is in quasi-TEM mode. The voltage and current intensity are expressed as follows:

$$i(z,t) = \oint \vec{j} \cdot \bar{z} dl \approx 2\pi r j(z,t).$$

(4.28)

$$q(z,t) = \oint \vec{\sigma} \cdot dl \approx 2\pi r \sigma(z,t).$$

(4.29)

Following the work of Maffucci et al. [50], combining Eqs. (4.28), (4.29), and (4.23), we obtain

$$\mathcal{E}_z = Ri + L_K \frac{\partial i}{\partial t} + \frac{1}{C_Q}\frac{\partial q}{\partial z},$$

(4.30)

where

$$R \equiv L_K \, \text{sgn}(l)v,$$

(4.31)

$$L_{\mathrm{K}} \equiv \frac{m}{2\pi re^2 n_0} \frac{1}{1+\alpha}, \tag{4.32}$$

$$C_{\mathrm{Q}} \equiv \frac{1}{L_{\mathrm{K}} u_e^2}. \tag{4.33}$$

In Eq. (4.32), L_{K} is the kinetic inductance per unit length of SWCNT because it is not a magnetic inductance and is related to the electron inertia. Substituting Eq. (4.27) into Eq. (4.32), we obtain

$$L_{\mathrm{K}} \equiv \frac{\pi\hbar}{4e^2 v_{\mathrm{F}}}, \tag{4.34}$$

In Eq. (4.33), C_{Q} is the quantum capacitance per unit length of SWCNT and is related to quantum pressure of the electron fluid. From Eqs. (4.19), (4.33), and (4.34), we obtain

$$\sqrt{\frac{L_{\mathrm{K}}}{C_{\mathrm{Q}}}} = \frac{\pi\hbar}{4e^2} \frac{1}{\sqrt{1-\alpha}} = \frac{R_0}{4} \frac{1}{\sqrt{1-\alpha}}, \tag{4.35}$$

where $R_0 = \pi\hbar/e^2 = 13$ kΩ is quantum resistance [45,50].

The parameter R in Eq. (4.30) is the resistance per unit length of metallic SWCNT. It depends on the electric field $\mathcal{E}_z$. By combining Eqs. (4.31), (4.14), and (4.2), we obtain

$$R(E_z) = \mathrm{sgn}(l)\frac{R_0}{4}\frac{v}{v_{\mathrm{F}}} = \mathrm{sgn}(l)\frac{R_0}{4}\frac{1}{\lambda}. \tag{4.36}$$

Finally, we need to consider the magnetic inductance and electric capacitance. The magnetic inductance per unit length of a perfect conductor on a ground plane is given by [63]

$$L_{\mathrm{M}} = \frac{\mu}{2\pi}\cosh^{-1}\left(\frac{h}{r}\right) \approx \frac{\mu}{2\pi}\ln\left(\frac{h}{2r}\right). \tag{4.37}$$

The electric capacitance per unit length of a perfect conductor on a ground plane is given by [63]

$$C_E = \frac{2\pi\varepsilon}{\cosh^{-1}(h/r)} \approx \frac{2\pi\varepsilon}{\ln(h/r)},$$ (4.38)

where h is the distance of SWCNT to the ground plane. Equations (4.37) and (4.38) are accurate enough for $h > 2r$.

The parameters R, L, and C, which are obtained above, are needed for building an equivalent circuit of metallic SWCNT interconnects. Typically, a nanotube radius is 1 nm. The oxide thickness over which the SWCNT is grown is between 100 Å and 1 μm. R, L, and C can be estimated as follows: $L_M \approx 1$ pH/μm, $L_K \approx 3.6$ nH/μm $\gg L_M$, which is close to measurement results [29], $C_Q \approx 90$ aF/μm and $C_E \approx 70$ aF/μm. The equivalent circuit of a metallic SWCNT interconnect is shown in Fig. 4.2.

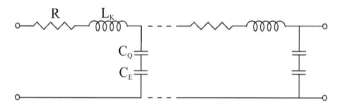

Figure 4.2   The equivalent circuit of a metallic SWCNT interconnect.

In the work of Burke [45], theoretical electrical contact resistance has been expressed as $\pi\hbar/4e^2$. Based on this theoretical contact resistance model, we have proposed an equivalent circuit model of a contact for experimental measurements [64] as shown in Fig. 4.3. In Fig. 4.3, R_{C1} is the theoretical contact resistance and R_{C2} and C_C are parasitic resistance and capacitance, respectively, which are to be measured during test.

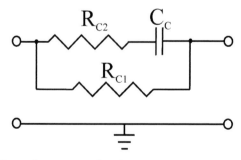

Figure 4.3   Equivalent circuit of an SWCNT contact.

4.5   Results and Discussion

The impedance, Z_{in}, with open circuit termination is one of the parameters used in characterizing the SWCNT transmission line behavior. For the transmission line shown in Fig. 4.2, Z_{in} can be calculated using

$$Z_{in} = Z_0 \frac{\cos(l\beta)}{i\sin(l\beta)},$$ (4.39)

where $\beta = u_e/f$ is the propagation constant. Z_0 is the characteristic impedance and is given by

$$Z_0 = \sqrt{\frac{R + j\omega L}{j\omega C}}.$$ (4.40)

Figure 4.4 shows the frequency dependence of Z_{in} for 0.1, 1, 10, and 100 µm lengths of SWCNTs. SWCNT interconnects show resonances at lower frequencies. Resonances of 0.1 and 1 µm SWCNT are shown in Fig. 4.4 as an inset. With increasing length, the resonance frequency range increases, thus limiting the use of SWCNT interconnects. However, considering the resonances range

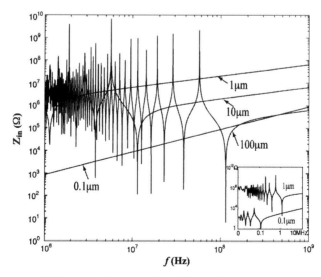

Figure 4.4 $|Z_{in}|$ versus frequency for different lengths of SWCNTs.

and applications, SWCNT interconnects are very promising. In short SWCNT interconnects (≤ 1 μm), resonance range is limited to 700 KHz, and thus, shorter length SWCNTs can be used for upper high-frequency (>1 MHz) applications. Long SWCNT interconnects (100 μm) exhibit no resonance beyond 1 GHz and thus can be still used for ultra-high-speed circuits. The input impedance, Z_{in}, of 0.1 μm length SWCNT is much smaller than that of 1 μm. This is due to the mean-free path of electrons in SWCNT, which is close to 1 μm and resistance, R, of 0.1 μm SWCNT is zero.

Figure 4.5 shows calculations of input impedance, Z_{in} for 10 μm-long SWCNT interconnect using the Lüttinger liquid model, two-dimensional (2D) fluid model and our one-dimension (1D) fluid model. The models differ in resonance frequency from each other within 20 MHz and difference in resistance value is less than 16 MΩ at 1 GHz.

The other important parameter is the scattering parameter, S, of the SWCNT transmission line in a two-port network. In calculating S parameters of an SWCNT, a comparison with Cu, widely used as interconnects, would be useful. In recent studies, performance of SWCNT and Cu interconnects has been compared [65–69] but the S parameters and group delay have not been studied. In the following, first we will discuss modeling of Cu interconnects in terms of R, L, and C parameters. As mentioned in Section 4.1, electron surface scattering and grain-boundary scattering contribute to increasing the resistivity of Cu in nanometer dimensions. The following equations describe the resistivity, ρ of Cu [70]:

$$\frac{\rho}{\rho_0} = 1 + \frac{3}{4}(1-p)\frac{\lambda_{Cu}}{w}, \tag{4.41}$$

$$\frac{\rho_0}{\rho} = 3\left\{\frac{1}{3} - \frac{\alpha}{2} + \alpha^2 - \alpha^3\left(1+\frac{1}{\alpha}\right)\right\}, \tag{4.42}$$

where $\alpha = \Gamma/[d(1-\Gamma)]$, ρ_0 is the bulk Cu resistivity, $p = 0.6$ is the fraction of electrons scattered at the surface, w is width of wire, $\lambda_{Cu} = 40$ nm is the mean free path of electrons in bulk Cu and d is the average grain size. $\Gamma = 0.5$ is the reflection coefficient describing the fraction of electrons that are not scattered by the potential barrier at a grain boundary. Based on the above models, at 22 nm node [71], Cu resistivity for minimum wire width increases to 5.7 μΩcm.

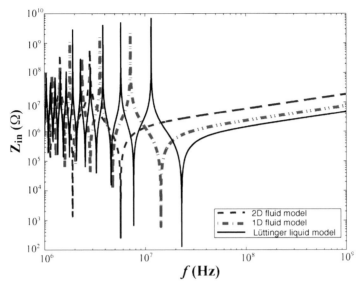

Figure 4.5 $|Z_{in}|$ versus frequency using different model for a 10 μm-long SWCNT.

The capacitance of Cu wire can be modeled by the following equation [72]:

$$C = \varepsilon \left\{ 1.15\frac{w}{h} + 2.80\left(\frac{t}{h}\right)^{0.222} + \left[0.06\frac{w}{h} + 1.66\left(\frac{t}{h}\right) - 0.14\left(\frac{t}{h}\right)^{0.222} \right]\left(\frac{h}{x}\right)^{1.34} \right\}, \qquad (4.43)$$

where ε is dielectric constant of the insulator, h is thickness of the insulator, t is the thickness of the wire and x is inter-wire spacing. At 22 nm node, typically, h = 216 nm, w = 27 nm, t = 54 nm and x = 27 nm.

The inductance of Cu wire can be modeled by the following equation [73]:

$$L = 2 \times 10^{-7} l_{Cu}\left(\ln\frac{2l_{Cu}}{w+t} + 0.5 + \frac{w+t}{3l_{Cu}} \right), \qquad (4.44)$$

where l_{Cu} is the length of the Cu wire.

Figure 4.6 shows the schematic of a two-port network used in study of S parameters. In Fig. 4.6, interconnect can be Cu or SWCNT. R_S is the terminating impedance. The terminating impedances of SWCNT and Cu interconnect lines used in S parameter studies are 3.2 kΩ and 50 Ω, respectively. In the case of SWCNT interconnect lines, the terminal impedance is equal to its contact resistance.

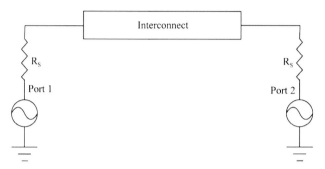

Figure 4.6   Schematic of a two-port network of interconnects.

Figures 4.7 and 4.8 show the performance of $|S_{21}|$ and $|S_{11}|$ parameters, respectively, for the SWCNT and Cu interconnects corresponding to 0.1, 1, 10, and 100 µm lengths. The calculations in Figs. 4.7 and 4.8 characterize the SWCNT transmission line with an ideal contact, $R_{\text{contact}} = \pi\hbar/4e^2$. Figure 4.7 shows the 3 dB bandwidth for $|S_{21}|$ for both SWCNT and Cu interconnects. The transmission efficiency of both the SWCNT and Cu interconnects decreases with increasing lengths. However, SWCNT transmission line has a larger 3 dB bandwidth when compared with Cu interconnects. Figure 4.7 also shows large $|S_{21}|$ for SWCNT interconnects of 1–100 µm lengths. $|S_{21}|$ for both SWCNT and Cu interconnects approaches to 0 dB for lengths less than 1 µm. Figure 4.7 also shows that the 3 dB bandwidth of short SWCNT interconnect lengths, 0.1 µm or less could exceed 1 THz.

Figure 4.8 shows $|S_{11}|$ for both SWCNT and Cu interconnects. $|S_{11}|$ of interconnects of 10–100 µm lengths show the values close to 0 dB at high frequencies. Whereas both SWCNT and Cu interconnects of 1 µm length show $|S_{11}|$ values close to −10 dB up to ~10 GHz then approaches to 0 dB. The performances in S_{11} parameter of SWCNT and Cu interconnects close to 0 dB for lengths more than 10 µm can be attributed to impedance mismatch at the input and output. SWCNT interconnect of 0.1 µm length shows noticeable suppression of $|S_{11}|$ parameter up to 100 GHz in comparison to equivalent length of Cu interconnect due to reactive impedance and then approaches to 0 dB beyond 1 THz.

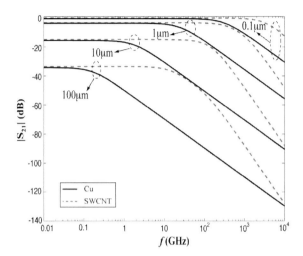

Figure 4.7 S_{21} (amplitude) versus frequency for different lengths SWCNT and Cu interconnects.

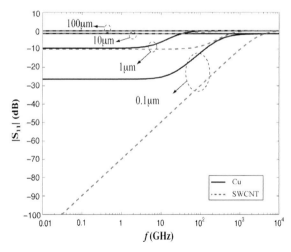

Figure 4.8 S_{11} (amplitude) versus frequency for different-length SWCNT and Cu interconnects.

Figures 4.9 and 4.10 show $|S_{21}|$ and $|S_{11}|$ parameters for 1 µm SWCNT interconnect lengths, respectively, calculated from the Lüttinger liquid model, 2D fluid model and our 1D fluid model. The insets in Figs. 4.9 and 4.10 show insignificant difference in $|S_{21}|$ and $|S_{11}|$ parameters calculated from three models. The 3 dB bandwidth obtained from Fig. 4.9 shows that our 1D fluid model gives nearly the same bandwidth as obtained from other two models.

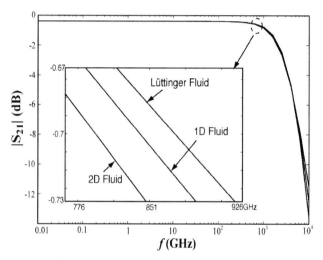

Figure 4.9 S_{21} (amplitude) versus frequency using different model for a 1 µm-long SWCNT.

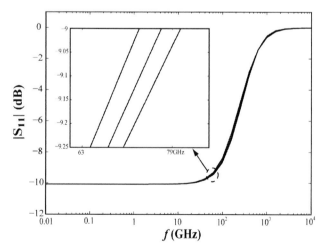

Figure 4.10 S_{11} (amplitude) versus frequency using different model for a 1 µm-long SWCNT.

Group delay is an import parameter. The variation from a constant value of group delay means signal distortion. Figure 4.11 shows the variation of group delay for both SWCNT and Cu interconnects with frequency. It is noticed from the Fig. 4.11 that shorter the length, lower is the delay for both SWCNT and Cu interconnects. SWCNT interconnects give group delay larger than the same length Cu interconnects but exhibits larger bandwidths. As an example, in Fig. 4.11, the group delay of 1 μm length Cu interconnect is larger than the 1 μm length SWCNT interconnect in the frequency range <100 GHz, while it becomes smaller in the frequency range >100 GHz. On the other hand, 1 μm SWCNT interconnect gives ~200 GHz bandwidth while the same length Cu interconnect gives ~10 GHz bandwidth. Figure 4.11 also shows significant improvement in group delay and bandwidth for longer-length SWCNTs as noticed for 100 μm SWCNT and Cu interconnect length.

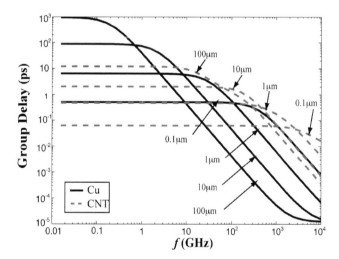

Figure 4.11   Group delay versus frequency for different lengths SWCNT and Cu interconnects.

We now consider the sound velocity, u_e in a metallic SWCNT. The relationship shown in Eq. (4.19) is very similar to that shown in the Lüttinger liquid model [45,46]. The ratio of Fermi velocity to the sound velocity in our model is about 0.5 and the ratio of Fermi velocity to the plasmon velocity is ~0.2–0.3 from experiments and theoretical estimates [74,75]. In the Lüttinger liquid model, the difference in Fermi velocity and plasmon velocity is due to electron–electron correlation, which is a quantum concept, referring to the interaction between electrons in a quantum system, the electronic structure of which is being considered. Therefore, the electron–electron correlation is distributed over the length of the SWCNT. The ratio is described by a dimensionless parameter, g. The parameter $g < 1$ for repulsive interactions, with small g meaning strong interactions, $g = 1$ describes the Fermi gas not a Fermi liquid and for attractive interactions, $g > 1$. Experiments in [52,53] show that there are significant electron–electron correlations in SWCNTs. From earlier measurements and theoretical calculations, the value of g is between 0.2–0.33 [52,53].

In 1D fluid model, the difference in Fermi velocity and sound velocity is due to electron–electron repulsive interaction, which is a classical concept, referring to the repulsive electrostatic forces only between electrons. Therefore, the electron–electron interaction is distributed over a short range. The ratio of Fermi velocity to sound velocity is described by a dimensionless parameter, $\sqrt{1 - \alpha} \approx 0.5$.

4.6 Summary

In this chapter, we have proposed a simple method to develop a transmission line model for metallic SWCNT interconnects using classical electrodynamics. The effective conducting electrons in carbon nanotubes are modeled as one-dimensional fluid considering electron–electron repulsive interactions. This method provides an equivalent circuit for analyzing the SWCNT interconnect as a transmission line. Damping effect is observed in SWCNTs due to its high resistance. It is observed in SWCNTs below 1 MHz for lengths less than 1 μm, and above 100 MHz for length longer than 100 μm. Thus, short length SWCNTs (<1 μm) can be used above 1 MHz. Damping thus limits the usable frequency bandwidth since it is dependent on the length. Calculations of group delays show that SWCNT interconnects can also be used above 200 GHz for short interconnects (<1 μm) and 10 GHz for long interconnect (>100 μm). Study of S parameters suggests consideration of impedance matching at the input and output to minimize losses due to reflections for longer SWCNT interconnects.

References

1. Saito, R., Dresselhaus, G., and Dresselhaus, M. S. (1998). *Physical Properties of Carbon Nanotubes*, (Imperial College Press, London, UK).
2. Dresselhaus, M. S., Dresselhaus, G., and Avouris, P. (2001) eds. *Carbon Nanotube: Synthesis, Properties, Properties and Applications* (Springer Verlag, Berlin Heidelberg New York).
3. Avouris, P., Appenzeller, J., Martel, R., and Wind, S. J. (2003). Carbon nanotube electronics, *Proc. IEEE*, **91**, 1772–1784.
4. Haselman, M., and Hauck, S. (2010). The future of integrated circuits: A survey of nanoelectronics, *Proc. IEEE*, **98**, 11–38.
5. Maffucci, A. (2009). Carbon nanotubes in nanopackaging applications, *IEEE Nanotechnol. Mag.*, **3**, 22–25.
6. Alam, N., Kureshi, A. K., Hasan, M., and Arslan, T. (2009). Carbon nanotube interconnects for low-power high-speed applications, *Proceedings of the IEEE International Symposium on Circuits and Systems* (*ISCAS*), pp. 2273–2276.
7. Cho, T. S., Lee, K.-J., Kong, J., and Chandrakasan, A. P. (2007). A low power carbon nanotube chemical sensor system, *Proceedings of the IEEE Custom Integrated Circuits Conference* (*CICC*), pp. 181–184.
8. Javey, A., Wang, Q., Ural, A., Li, Y., and Dai, H. (2002). Carbon nanotube transistor arrays for multistage complementary logic and ring oscillators, *Nano Lett.*, **2**, 929–932.
9. Frégonèse, S., Maneux, C., and Zimmer, T. (2009). Implementation of tunneling phenomena in a CNTFET compact model, *IEEE Trans. Electron Devices*, **56**, 2224–2231.
10. Tans, S. J., Verschueren, A. R. M., and Dekker, C. (1998). Room temperature transistor based on a single carbon nanotube, *Nature*, **393**, 49–52.
11. Wong, H. S. P. (2002). Field-effect transistors—from silicon MOSFETs to carbon nanotube FETs, *Proceedings of the 23rd International Conference on Microelectronics* (*MIEL 2002*), pp. 103–107.
12. Martel, R., Schmidt, T., Shea, H. R., Hertel, T., and Avouris, Ph. (1998). Single and multi-wall carbon nanotube field-effect transistors, *Appl. Phys. Lett.*, **73**, 2447–2449.
13. O'Connor, I., Liu, J., Gaffiot, F., Pregaldiny, F., Lallement, C., Maneux, C., Goguet, J., Fregonese, S., Zimmer, T., Anghel, L., Trong-Trinh, D., and Leveugle, R. (2007). CNTFET modeling and reconfigurable logic-circuit design, *IEEE Trans. Circuits Syst. I: Regular Papers*, **54**, 2365–2379.

14. Raychowdhury, A., Mukhopadhyay, S., and Roy, K. (2004). A circuit-compatible model of ballistic carbon nanotube field-effect transistors, *IEEE Trans. Comput.-Aided Des. Integrated Circuits Syst.*, **23**, 1411–1420.

15. Srivastava, A., Marulanda, J. M., Xu, Y., and Sharma, A. K. (2009). Current transport modeling of carbon nanotube field-effect transistors, *Phys. Status Solidi (A)*, **206**, 1569–1578.

16. Bachtold, A., Hadley, P., Nakanishi, T., and Dekker, C. (2001). Logic circuits with carbon nanotube transistors, *Science*, **294**, 1317–1320.

17. Javey, A., Wang, Q., Woong, K., and Dai, H. (2003). Advancements in complementary carbon nanotube field effect transistors, *IEDM Tech. Dig.*, **741**, 741–744.

18. Novak, J. P., Snow, E. S., Houser, E. J., Park, D., Stepnowski, J. L., and McGill, R. A. (2003). Nerve agent detection using networks of single-walled carbon nanotubes, *Appl. Phys. Lett.*, **83**, 4026–4028.

19. Snow, E. S., Perkins, F. K., Houser, E. J., Badescu, S. C., and Reinecke, T. L. (2005). Chemical detection with a single-walled carbon nanotube capacitor, *Science*, **307**, 1942–1945.

20. Qi, P., Vermesh, O., Grecu, M., Javey, A., Wang, Q., Dai, H., Peng, S., and Cho, K. J. (2003). Toward large arrays of multiplex functionalized carbon nanotube sensors for highly sensitive and selective molecular detection, *Nano Lett.*, **3**, 347–351.

21. Staii, C., Johnson, A. T., Chen, M., and Gelperin, A. (2005). DNA-decorated carbon nanotubes for chemical sensing, *Nano Lett.*, **5**, 1774–1778.

22. Star, A., Han, T.-R., Joshi, V., Gabrie, J.-C. P., and Grüner, G. (2004). Nanoelectronic carbon dioxide sensors, *Adv. Mater.*, **16**, 2049–2052.

23. Wilder, J. W. G., Venema, L. C., Rinzler, A. G., Smalley, R. E., and Dekker, C. (1998). Electronic structure of atomically resolved carbon nanotubes, *Nature*, **391**, 59–62.

24. Koo, K.-H., Kapur, P., and Saraswat, K. C. (2009). Compact performance models and comparisons for gigascale on-chip global interconnect technologies, *IEEE Trans. Electron Devices*, **56**, 1787–1798.

25. Ting, J.-H., Chiu, C.-C., and Huang, F.-Y. (2009). Carbon nanotube array vias for interconnect applications, *J. Vacuum Sci. Technol. B: Microelectron. Nanometer Struct.*, **27**, 1086–1092.

26. Li, H., Xu, C., Srivastava, N., and Banerjee, K. (2009). Carbon nanomaterials for next generation interconnect and passives: Physics, status and prospects, *IEEE Trans. Electron Devices*, **56**, 1799–1821.

27. Xu, T., Wang, Z., Miao, J., Chen, X., and Tan, C. M. (2007). Aligned carbon nanotubes for through-wafer interconnects, *Appl. Phys. Lett.*, **91**, 042108-1 to 042108-3.

28. Chiariello, A. G., Maffucci, A., and Miano, G. (2009). Signal integrity analysis of carbon nanotube on-chip interconnects, *Proceedings of the IEEE Workshop on Signal Propagation on Interconnects* (*SPI 2009*), pp. 1–4.

29. Plombon, J. J., O'Brien, K. P., Gstrein, F., Dubin, V. M., and Jiao, Y. (2007). High-frequency electrical properties of individual and bundled carbon nanotubes, *Appl. Phys. Lett.*, **90**, 063106-3.

30. Yao, Z., Kane, C. L., and Dekker, C. (2000). High-field electrical transport in single-wall carbon nanotubes, *Phys. Rev. Lett.*, **84**, 2941–2944.

31. Nougaret, L., Dambrine, G., Lepilliet, S., Happy, H., Chimot, N., Derycke, V., and Bourgoin, J. P. (2010). Gigahertz characterization of a single carbon nanotube, *Appl. Phys. Lett.*, **96**, 042109-1 to 042109-3.

32. Sarto, M. S., and Tamburrano, A. (2010). Single-conductor transmission-line model of multiwall carbon nanotubes, *IEEE Trans. Nanotechnol.*, **9**, 82–92.

33. Novoselov, K. S., Geim, A. K., Morozov, S. V., Jiang, D., Zhang, Y., Dubonos, S. V., Grigorieva, I. V., and Firsov, A. A. (2004). Electric field effect in atomically thin carbon films, *Science*, **306**, 666–669.

34. Naeemi A., and Meindl, J. D. (2009). Compact physics-based circuit models for graphene nanoribbon interconnects, *IEEE Trans. Electron Devices*, **56**, 1822–1833.

35. Ryu, C., Kwon, K.-W., Loke, A. L. S., Lee, H., Nogami, T., Dubin, V. M., Kavari, R. A., Ray, G. W., and Wong, S. S. (1999). Microstructure and reliability of copper interconnects, *IEEE Trans. Electron Devices*, **46**, 1113–1120.

36. Miller, D. A. B. (2000). Optical interconnects to silicon, *IEEE J. Selected Top. Quantum Electron.*, **6**, 1312–1317.

37. Chen, G., Chen, H., Haurylau, M., Nelson, N. A., Albonesi, D. H., Fauchet, P. M., and Friedman, E. G. (2007). Predictions of CMOS compatible on-chip optical interconnect, *VLSI J.*, **40**, 434–446.

38. Kapur P., and Saraswat, K. C. (2002). Comparisons between electrical and optical interconnects for on-chip signaling, *Proceedings of the IEEE 2002 International Interconnect Technology Conference*, pp. 89–91.

39. Chiariello, A. G., Miano, G., and Maffucci, A. (2009). Carbon nanotube bundles as nanoscale chip to package interconnects, *Proceedings of the 9th IEEE Conference on Nanotechnology* (*IEEE-NANO 2009*), pp. 58–61.

40. Stan, M. R., Unluer, D., Ghosh, A., and Tseng, F. (2009). Graphene devices, interconnect and circuits—challenges and opportunities, *Proceedings of the IEEE International Symposium on Circuits and Systems* (*ISCAS 2009*), pp. 69–72.

41. Srivastava, N., Li, H., Kreupl, F., and Banerjee, K. (2009). On the applicability of single-walled carbon nanotubes as VLSI interconnects, *IEEE Trans. Nanotechnol.*, **8**, 542–559.

42. Patel-Predd, P. (2008). Update: Carbon-nanotube wiring gets real, *IEEE Spectr.*, **45**, 14.

43. Iijima, S. (1991). Helical microtubules of graphitic carbon, *Nature*, **354**, 56–58.

44. Fisher, M. P. A., and Glazman, L. I. (1996). Transport in one-dimensional Luttinger liquid. In: *Mesoscopic Electron Transport* (ed. Sohan, L. L., Kouwenhoven, L. P., and Schoen, G.), NATO ASI Series, **345**, p. 331.

45. Burke, P. J. (2003). An RF circuit model for carbon nanotubes, *IEEE Trans. Nanotechnol.*, **2,** 55–58.

46. Burke, P. J. (2002). Luttinger liquid theory as a model of the gigahertz electrical properties of carbon nanotubes, *IEEE Trans. Nanotechnol.*, **1**, 129–144.

47. Salahuddin, S., Lundstrom, M., and Datta, S. (2005). Transport effects on signal propagation in Quantum wires, *IEEE Trans. Electron Devices*, **52**, 1734–1742.

48. Fetter, A. L. (1973). Electrodynamics of a layered electron gas. I. single layer, *Ann. Phys.*, **81**, 367–393.

49. Fetter, A. L. (1974). Electrodynamics of a layered electron gas. II. Periodic array, *Ann. Phys.*, **88**, 1–25.

50. Maffucci, A., Miano, G., and Villone, F. (2008). A transmission line model for metallic carbon nanotube interconnects, *Int. J. Circuit Theory Appl.*, **36**, 31–51.

51. Kane, C., Balents, L., and Fisher, M. P. A. (1997). Coulomb interactions and mesoscopic effects in carbon nanotubes, *Phys. Rev. Lett.*, **79**, 5086–5089.

52. Bockrath, M., Cobden, D. H., Lu, J., Rinzler, A. G., Smalley, R. E., Balents, L., and McEuen, P. L. (1999). Luttinger-liquid behavior in carbon nanotubes, *Nature*, **397**, 598–601.

53. Ishii, H., Kataura, H., Shiozawa, H., Yoshioka, H., Otsubo, H., Takayama, Y., Miyahara, T., Suzuki, S., Achiba, Y., Nakatake, M., Narimura, T., Higashiguchi, M., Shimada, K., Namatame, H., and Taniguchi, M. (2003).

Direct observation of Tomonaga-Luttinger-liquid state in carbon nanotubes at low temperatures, *Nature*, **426**, 540–544.

54. Civalleri, P. P., Gilli, M., and Bonnin, M. (2007). Equivalent circuits for two-state quantum systems, *Int. J. Circ. Theor. Appl.*, **35**, 265–280.

55. Xu Y., and Srivastava, A. (2010). A model for carbon nanotube interconnects, *Int. J. Theor. Appl.*, **38**, 559–575.

56. Miano, G., and Villone, F. (2006). An integral formulation for the electrodynamics of metallic carbon nanotubes based on a fluid model (2006). *IEEE Trans. Antennas Propagation*, **54**, 2713–2724.

57. Chiariello, A. G., Maffucci, A., Miano, G., Villone, F., and Zamboni, W. (2006). Metallic carbon nanotube interconnects, part I: A fluid model and a 3D integral formulation, *Proceedings of the IEEE Workshop on Signal Propagation on Interconnects*, pp. 181–184.

58. Shadowitz, A. (1988). *The Electromagnetic Field* (Courier Dover Publications).

59. Batchelor, G. K. (1967). *An Introduction to Fluid Dynamics* (Cambridge University Press, London, UK).

60. Park, J.-Y., Rosenblatt, S., Yaish, Y., Sazonova, V., Ustunel, H., Braig, S., Arias, T. A., Brouwer, P. W., and McEuen, P. L. (2004). Electron-phonon scattering in metallic single-walled carbon nanotubes, *Nano Lett.*, **4**, 517–520.

61. McEuen, P. L., Fuhrer, M. S., and Park, H. (2002). Single-walled carbon nanotube electronics, *IEEE Trans. Nanotechnol.*, **1**, 78–85.

62. Slepyan, G. Y., Maksimenko, S. A., Lakhtakia, A., Yevtushenko, O., and Gusakov, A. V. (1999). Electrodynamics of carbon nanotubes: Dynamic conductivity, impedance boundary conditions, and surface wave propagation, *Phys. Rev. B*, **60**, 17136–17149.

63. Ramo, S., Whinnery, J. R., and Duzer, T. V. (1994). *Fields and Waves in Communication Electronics*, 3rd ed. (John-Wiley & Sons, Inc.).

64. Xu, Y., and Srivastava, A. (2007). A two-port network model of CNT-FET for RF characterization, *Proceedings of the 50th IEEE Midwest Symposium on Circuits and Systems* (*MWSCAS 2007*), pp. 626–629.

65. Naeemi, A., and Meindl, J. D. (2008). Performance modeling for single- and multi-wall carbon nanotubes as signal and power interconnects in gigascale systems, *IEEE Trans. Electron Devices*, **55**, 2574–2582.

66. Cho, H., Koo, K.-H., Kapur, P., and Saraswat, K. C. (2007). The delay, energy, and bandwidth comparisons between copper, carbon nanotube, and optical interconnects for local and global wiring

application, *Proceedings of the IEEE International Interconnect Technology Conference*, pp. 135–137.

67. Koo, K.-H., Cho, H., Kapur, P., and Saraswat, K. C. (2007). Performance comparisons between carbon nanotubes, optical, and Cu for future high-performance on-chip interconnect applications, *IEEE Trans. Electron Devices*, **54**, 3206–3215.

68. Nieuwoudt, A., and Massoud, Y. (2008). On the optimal design, performance, and reliability of future carbon nanotube-based interconnect solutions, *IEEE Trans. Electron Devices*, **55**, 2097–2110.

69. Naeemi A., and Meindl, J. D. (2007). Design and performance modeling for single-walled carbon nanotube as local, semiglobal and global interconnects in gigascale integrated systems, *IEEE Trans. Electron Devices*, **54**, 26–37.

70. Steinhögl, W., Schindler, G., Steinlesberger, G., and Engelhardt, M. (2002). Size-dependent resistivity of metallic wires in the mesoscopic range, *Phys. Rev. B*, **66**, 075414-1 to 075414-4.

71. International Technology Roadmap for Semiconductors. (2007). Available: http://www.itrs.net/Links/2007ITRS/Home2007.htm.

72. Sakurai, T., and Tamaru, K. (1983). Simple formulas for two- and three-dimensional capacitances, *IEEE Trans. Electron Devices*, **30**, 183–185.

73. Yue, C. P., and Wong, S. S. (2000). Physical modeling of spiral inductors on silicon, *IEEE Trans. Electron Devices*, **47**, 560–568.

74. Bockrath, M., Cobden, D. H., McEuen, P. L., Chopra, N. G., Zettl, A., Thess, A., and Smalley, R. E. (1997). Single-electron transport in ropes of carbon nanotubes, *Science*, **275**, 1922–1925.

75. Tans, S. J., Devoret, M. H., Dai, H., Thess, A., Smalley, R. E., Geerligs, L. J., and Dekker, C. (1997). Individual single-wall carbon nanotubes as quantum wires, *Nature*, **386**, 474–477.

Chapter 5

Multi-Walled and Bundle of Single-Walled Carbon Nanotube Interconnection

5.1 Introduction

A good amount of research has been conducted in modeling single-walled carbon nanotubes (SWCNT), SWCNT bundle, and multi-walled carbon nanotube (MWCNT) interconnects [1–10]. An SWCNT has very large contact resistance [11–13], which limits its application as an interconnect for next-generation integrated circuits. On the other hand, MWCNT and CNT bundle give low contact resistance when used as the circuit interconnects [14–18]. One-dimensional fluid model, which has been applied in modeling of an SWCNT interconnect derived in Chapter 4 can also be extended in modeling multi-walled and bundle of SWCNT interconnects [19], which is described in the following sections.

5.2 MWCNT Interconnection Modeling

Multi-walled carbon nanotubes have diameters in a wide range of a few to hundreds of nanometers. It has been shown that all shells of an MWCNT can conduct if they are properly connected to contacts [15–17] and the contact resistance could reach tens of ohms, a much

Carbon-Based Electronics: Transistors and Interconnects at the Nanoscale
Ashok Srivastava, Jose Mauricio Marulanda, Yao Xu, and Ashwani K. Sharma
Copyright © 2015 Pan Stanford Publishing Pte. Ltd.
ISBN 978-981-4613-10-1 (Hardcover), 978-981-4613-11-8 (eBook)
www.panstanford.com

lower value than that of an SWCNT. Naeemi et al. [8] have shown that MWCNTs can have conductivities several times larger than that of Cu or SWCNT bundles for long-length interconnects.

The number of shells in MWCNTs varies. The spacing between shells in an MWCNT corresponds to van der Waals distance between graphene layers in graphite, $\delta \approx 0.34$ nm [20]. The number of metallic shells in an MWCNT can be calculated as follows:

$$M = \beta \left[1 + \frac{D_1 - D_N}{2\delta} \right],$$
(5.1)

where D_1 and D_N are the outermost and innermost shell diameters, respectively. The square bracket term is a floor function and the factor β is the ratio of metallic shells to total shells in an MWCNT. Statistically, one-third of the shells are going to be metallic and the rest semiconductor for $D_1 \leq 10$ nm [20,21]. For $D_1 > 10$ nm, β increases due the interaction between adjacent shells for the MWCNT [20].

In one-dimensional fluid model [22], we regard the graphene sheet that is rolled to form a CNT is to be infinitesimally thin. The conduction electrons are then distributed over the lateral surface of the CNT cylinder shell and electrons are embedded in a rigid uniform positive charge background with a uniform surface number density. Thus, the motion of electrons is confined to the surface. Furthermore, electrical charge neutrality requires that in equilibrium, the conduction electron charge density precisely cancel with that of the background positive ions. Since the van der Waals force between the carbon atoms in different shells in MWCNTs is negligible compared to valence band between the carbon atoms in the same shell [23], the one-dimensional fluid model described by Eq. (4.9) in Chapter 4 can be applied to each shell of the MWCNT with modification because the electron–electron interaction in the MWCNT is different from that in the SWCNT, which means the parameter α needs to be recalculated.

In addition, two assumptions are made: The electrons can only move along the z-axis; all other fluid variables, such as the tangential component of the electric field to the nanotube surface, are almost uniform in the cross section plane of the shells in MWCNT. These two assumptions are valid if both the nanotube length and the smallest wavelength of the electromagnetic field are much greater than the nanotube radius [24,25]. Furthermore, the analysis assumes room temperature operation and on distribution of elections in other directions than in z-direction for the MWCNT and SWCNT bundle interconnections.

We assume that the velocity of these electrons equals the Fermi velocity. As a result, the kinetic energy is given by

$$E_K = 4M \times \frac{1}{2}mv_F^2 \approx 7M \text{ eV} \tag{5.2}$$

There are two channels in each shell of the MWCNT and two different spin electrons in each channel. So we consider that there are four electrons at the same point in each shell of the MWCNT. We can then calculate the potential energy by moving these 4M electrons from ∞ to the same point of the MWCNT. We first consider moving every four electrons into one shell of the MWCNT. The potential energy can be obtained as follows [22,26]:

$$E_P = \sum_{j=1}^{M}\left[\sum_{i=2}^{4}(i-1)\frac{e^2}{2\pi\varepsilon_0}\frac{1}{d_j}\right] = 6M\frac{e^2}{2\pi\varepsilon_0}\sum_{j=1}^{M}\frac{1}{d_j}, \tag{5.3}$$

where d_j is diameter of shell number j.

We then consider moving all shells from ∞ to adjacent shells to construct an MWCNT. The potential energy can be calculated using following equation:

$$E_P = 6M\frac{e^2}{2\pi\varepsilon_0}\sum_{j=1}^{M}\frac{1}{d_j} + \sum_{j=2}^{M}4(j-1)\times\frac{4e^2}{2\pi\varepsilon_0}\frac{1}{d_j} = 6M\frac{e^2}{2\pi\varepsilon_0}\sum_{j=1}^{M}\frac{1}{d_j} + \frac{16e^2}{2\pi\varepsilon_0}\sum_{j=2}^{M}\frac{j-1}{d_j} \tag{5.4}$$

The parameter α for the MWCNT can be calculated using Eq. (4.6). For example, if $D_1 = 10$ nm, $D_N = 1$ nm and $\beta = 1$ then $\alpha \approx 0.99$.

Following the derivation in [22], we can obtain an equation for each shell in an MWCNT:

$$\mathcal{E} = Ri + L_K\frac{\partial i}{\partial t} + \frac{1}{C_Q}\frac{\partial q}{\partial z}, \tag{5.5}$$

where $R \equiv L_K \text{ sgn }(I)v$ is the resistance of each shell in an MWCNT per unit length. $L_K \equiv \pi\hbar/4e^2v_F$ is the kinetic inductance per unit length of each shell. $CQ \equiv 1/L_K u_e^2$ is the quantum capacitance per unit length of each shell. $u_e = v_F/\sqrt{1-\alpha}$ is the thermodynamic speed of sound of the electron fluid under a neutral environment.

The magnetic inductance per unit length of each shell can also be calculated using Eq. (4.37). In an SWCNT, magnetic inductance is neglected compared with kinetic inductance; therefore, it can also be neglected in each shell of an MWCNT.

The outermost shell shields inner shells from the ground plane; therefore, the electrostatic capacitance, C_E, does not exist in inner shells. However, there exists electrostatic capacitance, C_S, between the neighboring metallic shells and its value is given by [26,27]

$$C_S = \frac{2\pi\varepsilon_0}{\ln(D_i/D_j)},$$
(5.6)

where ε_0 is the permittivity of vacuum, D_i and D_j are the diameters of the i-th and j-th metallic shells, respectively and $I < j$.

We assume that the outermost shell is metallic. In a recent work [27], we have derived an equivalent circuit of a metallic MWCNT interconnect as shown in Fig. 5.1. It is simplified as shown in Fig. 5.2 by considering that the RLC parts of all inner shells are identical. If we assume that there are no variation in distributed parameters, R and L_K, then R and L_K are same for each shell. The potential across components of each shell in an MWCNT is equal. As a result, a simplified equivalent circuit of an MWCNT interconnect can be derived as shown as Fig. 5.3. R_C in Figs. 5.1–5.3 is the contact resistance and its ideal quantum value is 3.2 kΩ per shell [12].

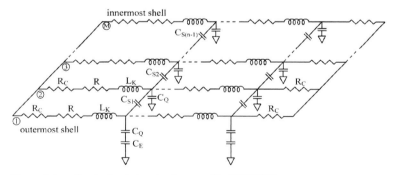

Figure 5.1   Equivalent circuit of a metallic MWCNT interconnect.

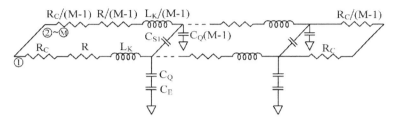

Figure 5.2 Simplified equivalent circuit of a metallic MWCNT interconnect.

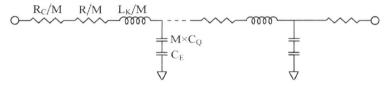

Figure 5.3 Simple equivalent circuit model of a metallic MWCNT interconnect.

The values of C_E and C_Q are on same order. C_Q of all metallic shells is in parallel and then serial with C_E. As a result, C_Q can be neglected if M is large. Therefore, capacitance of an MWCNT interconnect is smaller than that of SWCNT. In addition, the resistance and inductance of all metallic shells are parallel and M times smaller than that of an SWCNT.

5.3 SWCNT Bundle Interconnection Modeling

Carbon nanotubes can also be fabricated as a bundle, which means CNTs in a bundle are parallel to each other. The spacing between carbon nanotubes in the bundle is due to the van der Waals forces between the atoms of adjacent nanotubes [28]. One of the most critical challenges in realizing high-performance SWCNT-based interconnects is controlling the proportion of metallic nanotubes in the bundle. Current SWCNT fabrication techniques cannot effectively control the chirality of the nanotubes in the bundle [29,30]. Therefore, SWCNT bundles have metallic nanotubes that are randomly distributed within the bundle. Avouris et al. [29] and Liebau et al. [30] have shown that metallic nanotubes are distributed with a probability $\beta = 1/3$ in a growth process. The proportion of metallic nanotubes can, however, be potentially increased using techniques introduced in [31,32].

Figure 5.4 shows the cross section of an SWCNT bundle. In Fig. 5.4, d is diameter of SWCNT, δ = 0.34 nm is the spacing between the SWCNT in the bundle and corresponds to the van der Waals distance between graphene layers in graphite. The distance between the adjacent SWCNT is $d_b = \delta + d$. Since van der Waals forces between carbon atoms in adjacent SWCNTs are negligible compared to valence band between carbon atoms in the SWCNT [23], influence of adjacent SWCNTs on transport of electrons in SWCNT can be considered to be very small. Therefore, the one-dimensional fluid model described by Eq. (4.9) can be applied to the each SWCNT in the bundle with some modification. The electron–electron interaction in SWCNT bundle is different from that in an SWCNT, the parameter α needs to be recalculated to account for this effect.

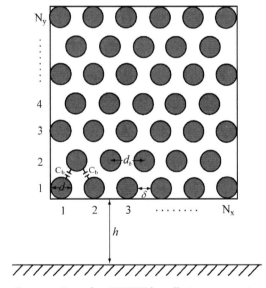

Figure 5.4   Cross section of an SWCNT bundle interconnect.

Considering one of the SWCNTs in the bundle, we assume that the electrons in this SWCNT will only be affected by the electrons in the adjacent metallic SWCNTs and semiconducting SWCNTs have no effect on the conductance of the bundle. Here we assume that SWCNTs in the bundle are insulated from each other and no conduction takes place between them. To calculate the potential energy, we first consider the potential energy of each SWCNT and then consider moving SWCNT adjacent to each other to construct an SWCNT bundle. Average potential energy of electrons in an SWCNT bundle can then be described by the following equation:

$$E_P = 6\frac{e^2}{2\pi\varepsilon_0}\frac{1}{d} + \sum_{i=1}^{\Gamma}\frac{16e^2}{2\pi\varepsilon_0}\frac{1}{d_b}, \tag{5.7}$$

where Γ is the average number of metallic SWCNTs neighboring to a single SWCNT. As shown in Fig. 5.4, e.g., the number of SWCNTs neighboring to the corner SWCNT is 2, the number of SWCNTs neighboring to the edge SWCNT is 4, and the number of SWCNTs neighboring to the inside SWCNT is 6.

Therefore, $\Gamma = \beta \times \left[\dfrac{6N_xN_y - 4N_x - 4N_y - 2\left[N_y/2\right]}{N_xN_y - 2\left[N_y/2\right]}\right]$, where the

square brackets denote the floor function.

The kinetic energy of the electrons in an SWCNT is calculated as follows [22]:

$$E_K = 4 \times \frac{1}{2}m_e v_F^2 \approx 7 \text{ eV}, \tag{5.8}$$

Now the parameter α for SWCNT bundle can be calculated using Eq. (4.6) in Chapter 4. Total number of metallic SWCNTs is $N = \beta(N_xN_y - [N_y/2])$. Following the derivation in [22], we obtain an equation for each single SWCNT in a bundle:

$$E = Ri + L_K\frac{\partial i}{\partial t} + \frac{1}{C_Q}\frac{\partial q}{\partial z}, \tag{5.9}$$

where $R \equiv L_K \text{ sgn }(I)v$ is the resistance per unit length of an SWCNT in an SWCNT bundle.

In Eq. (5.9), $L_K \equiv \pi\hbar/4e^2 v_F$ is the kinetic inductance per unit length of an SWCNT in a bundle and $C_Q \equiv 1/L_K u_e^2$ is the quantum capacitance per unit length of an SWCNT in a bundle. The relation $u_e = v_F/\sqrt{1-\alpha}$ is the thermodynamic speed of sound of the electron fluid under a neutral environment.

The magnetic inductance per unit length of each SWCNT can also be calculated using Eq. (3.37). In a single SWCNT, magnetic inductance is neglected compared with the kinetic inductance. It can also be neglected in each SWCNT of a bundle since magnetic inductance is comparable to the kinetic conductance when the number of SWCNTs is above 4000 in a bundle [4], while this number is only about 500 for the bundle with 22 nm × 44 nm size. This size of the bundle corresponds to interconnections in 22 nm technology.

The SWCNTs at the bottom level shield upper levels SWCNT from the ground plane. Therefore, the electrostatic capacitance, C_E, does not exist in the upper SWCNTs. However, there exists electrostatic capacitance per unit length, C_b, between the neighboring metallic SWCNTs and its value can be calculated as follows [26]:

$$C_b = \pi \varepsilon_0 \left/ \ln\left(\frac{d_b}{d} + \sqrt{\left(\frac{d_b}{d}\right)^2 - 1}\right)\right. . \qquad (5.10)$$

Figure 5.5 shows the equivalent circuit of an SWCNT bundle interconnect [33] where N_a is the number of upper level SWCNTs and N_b is the number of bottom level SWCNTs. For an SWCNT bundle, we assume that all SWCNTs in the bundle are identical and each SWCNT has the same potential across it [34,35]. The circuit can be further simplified as shown in Fig. 5.6. The capacitance, C_b has no effect on the circuit behavior and $\beta N_x \times C_E$ can be regarded as an electrostatic capacitance between SWCNT bundle and the ground plane.

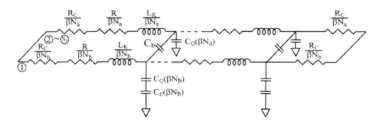

Figure 5.5   Equivalent circuit of an SWCNT bundle interconnect.

The values of C_E and C_Q are nearly the same in magnitude. The C_Q values of all metallic SWCNTs are in parallel and then serial with C_E; as a result, C_Q can be neglected if N is large. Therefore, capacitance of the SWCNT bundle interconnects is smaller than that of an SWCNT. In addition, the resistance and inductance of all metallic SWCNTs are in parallel in the bundle and N times smaller than that of an SWCNT.

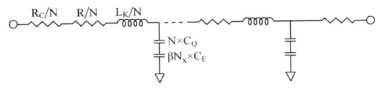

Figure 5.6 Simplified equivalent circuit of an SWCNT bundle interconnect.

5.4 Performance of MWCNT and Bundles of SWCNT Interconnects

In Section 5.1, we have extended one-dimensional fluid model for the modeling of MWCNT interconnects. To validate the model of MWCNT interconnect, we have compared the calculated resistance of MWCNT interconnect with the corresponding measured resistance from the work of Nihei et al. [15] and Li et al. [16]. The results of the comparison of calculated and measured resistances are summarized in Table 5.1.

Table 5.1 A comparison of calculated and measured resistances of MWCNT interconnects

| References | MWCNT physical parameters | | | MWCNT resistance (kΩ) | | |
	Length (μm)	D_1 (nm)	D_n (nm)	l_{mfp} (μm)	Measured	Model
Nihei et al. [15]	2	10	3.88	<1	1.60	1.90
Li et al. [16]	25	100	50	>25	0.035	0.042

As mentioned in Chapter 4, CNT interconnects have great potential in extending the operation of circuits to higher speeds and frequencies. For applications requiring high frequencies where newer interconnect technologies and materials for interconnect are being explored, it is important to study two-port scattering parameters. Slepyan et al. [36] have conducted studies on scattering of electromagnetic waves by a semi-infinite CNT in optical regime. Here, we focus on studying two-port network scattering (S) parameters by CNT for interconnect applications on a chip.

We have used the schematic shown in Fig. 4.6 in Chapter 4 to study S-parameters and have utilized the process parameters from the 2016 node (22 nm technology), assumed a 22 nm diameter of MWCNT of 22 nm width and 44 nm thickness of SWCNT bundle [37]. If we assume the diameter of the innermost shell in an MWCNT is to be 1 nm, then there are nearly 30 shells in 22 nm MWCNT. If we assume diameter of an SWCNT in a bundle is to be 1 nm then there are nearly 500 SWCNTs in 22 nm (width) × 44 nm (thickness) bundle following Fig. 5.4. The resistivity and capacitance of Cu were taken from ITRS 2007 [37]. The inductance of Cu wire can be modeled by the Eq. (4.44).

An SWCNT has very large contact resistance [11–13] when used as interconnects, which limits its applications as interconnects for next-generation integrated circuits. On the other hand, MWCNT and CNT bundle give low contact resistance when used as circuit interconnects [14–18]. Contact resistance in MWCNT and SWCNT bundle, however, will depend on the number of shells or SWCNTs being metallic. Close et al. [38–40] have demonstrated experimentally that the MWCNT can function as an interconnect wire on a chip and successfully transmit GHz digital signals from one transistor to another. In Table 5.2, modeling parameters of MWCNT are compared with the equivalent model parameters from the quantum theory [8,9]. The difference is about 20%. However, for large diameter, such as 100 nm, the difference reaches about 60%. According to the quantum theory, the number of channels increases significantly for large values of radius [34] and the semiconducting shells start contributing significantly to the number of conducting channels since their axial conductivity increases with increasing radius [8,41]. Our semi-classical one-dimensional fluid model assumes that the number of conducting channels in a single metallic CNT shell is fixed, 2, and 0 for semiconducting shells. Therefore, the difference in values of parameters between our model and quantum theory increases with the increase in diameter of MWCNTs. On the other hand, the parameter α decreases with the increase in diameter and quantum capacitance increases with the increase in diameter, which is consistent with the quantum theory [8,9]. As a result, the difference in values of parameters calculated from our model and quantum theory are not very large for small-diameter MWCNTs. The electrostatic capacitance is dependent on the geometry of the structure; it is, thus, considered the same for one-dimensional fluid model and quantum model [9].

Table 5.2   A comparison of MWCNT interconnect model parameters

MWCNT diameter (nm)	R_C (cal) kΩ	R_C [9] kΩ	R (cal) kΩ/μm	R [9] kΩ/μm	L_K (cal) nH/μm	L_K [9] nH/μm	C_Q (cal) aF/μm	C_Q [181] aF/μm
18	0.81	1.05	0.81	1.05	1.00	1.31	1280	1160
20	0.65	0.78	0.65	0.78	0.80	0.97	1600	1566
22	0.65	0.72	0.65	0.72	0.80	0.90	1600	1682
25	0.54	0.55	0.54	0.55	0.67	0.68	1920	2228
28	0.46	0.43	0.46	0.43	0.57	0.53	2240	2844
32	0.40	0.34	0.40	0.34	0.50	0.42	2560	3622
90	0.16	0.06	0.16	0.06	0.21	0.08	11080	19208
100	0.11	0.04	0.11	0.04	0.17	0.05	17680	29845

Figures 5.7 and 5.8 show S_{21} and S_{11} parameters and comparison with the corresponding S-parameters for MWCNTs calculated using model parameters from the work of Li et al. [9]. The dimensions used in comparison correspond to 18, 22, and 32 nm diameters of the outermost shells of MWCNTs, which also correspond to nm technologies. The length of MWCNTs used in calculations is 10 μm. Terminal impedance is set equal to contact resistance and $D_1/D_N = 2$ and $\beta = 1/3$. The parameters S_{21} and S_{11} in both models differ by about 6% corresponding to 18 nm diameter and it is less than 6% for 22 and 32 nm diameters. The phase difference is negligible within the 3 dB bandwidths. It can, thus, be stated that the one-dimensional fluid model can be easily used in studying the performance of MWCNT interconnects.

Figures 5.9a,b show S_{21} and S_{11} parameters of MWCNT and SWCNT bundles and Cu interconnects of lengths corresponding to ballistic transport (1 μm), local interconnection (10 and 100 μm) and global interconnection (500 μm). For comparison, we choose $\beta = 1/3$ and 50 Ω terminal impedance, which is a typical impedance for high-frequency transmission lines. For the MWCNT and SWCNT bundles, the electrostatic capacitance depends on the geometry of the structure and is approximately equal to that of Cu interconnects [9,35,42].

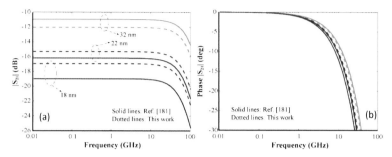

Figure 5.7   Comparison of S_{21} from our model and Li et al. model [9] for MWCNT interconnects: (a) amplitude and (b) phase.

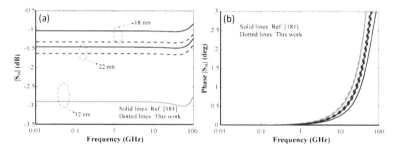

Figure 5.8   Comparison of S_{11} from our model and Li et al. model [9] for MWCNT interconnects: (a) amplitude and (b) phase.

Figure 5.9a shows the 3 dB bandwidths for both the CNT and Cu interconnects. The transmission efficiency of both the CNT and Cu interconnects decreases with increasing lengths. However, the Cu interconnect has a larger 3 dB bandwidth in comparison with CNT interconnects. This is because Cu has smaller inductance compared with CNT interconnects. It should also be noticed that the short length CNT interconnects still have over 100 GHz 3 dB bandwidth. Figure 5.9a also shows large S_{21} for SWCNT bundle and MWCNT interconnects than that of the Cu interconnect. This is because SWCNT bundle and MWCNT have much smaller resistances. Furthermore, an SWCNT bundle has more connection channels than MWCNT, it has larger 3 dB bandwidth and S_{21} value, which means larger transmission efficiency. In Fig. 5.9b, for S_{11}, parameters at frequencies less than 100 GHz, Cu interconnect has the largest reflection losses while SWCNT bundle interconnect has the least reflection losses. The results show that SWCNT bundle interconnect has better performance than the MWCNT interconnect. This can be explained that the number of SWCNTs in the bundle is larger than that of shells in the MWCNT of the same size. It can be shown that for 22 nm width of SWCNT bundle and MWCNT interconnects calculated number of SWCNTs in a bundle from $N = \beta(N_x N_y - [N_y/2])$ and the number of shells in an MWCNT from Eq. (5.1) are approximately 500 and 10, respectively. This means that there are more conducting channels in the bundle according to one-dimensional fluid model.

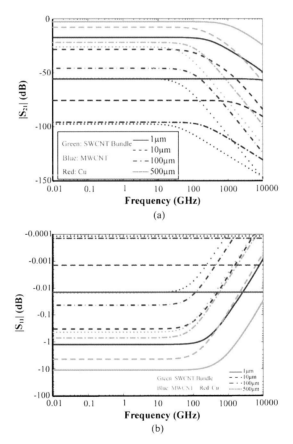

(a)

(b)

Figure 5.9 Calculated S-parameters of different interconnects: (a) S_{21} (amplitude) and (b) S_{11} (amplitude).

Figure 5.10 shows the CNT-FET inverter pair at 1 V supply voltage. The interconnection can be Cu or MWCNT or SWCNT bundle. The delay analysis includes the CNT-FET models developed by

Srivastava et al. [43] and our dynamic models reported in [44]. In this work, Verilog-AMS is used to describe CNT-FET static and dynamic models and simulated CNT-FET circuits in Cadence/Spectre.

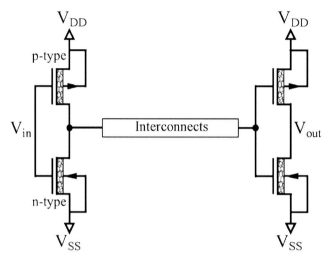

Figure 5.10   Inverter pair with interconnects.

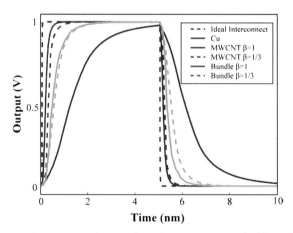

Figure 5.11   Output waveforms of an inverter pair with 10 μm length of different interconnect wires.

Figure 5.11 shows transient response of a CNT-FET inverter interconnected with 10 μm-long MWCNT and SWCNT bundle interconnection wires. Figure 5.11 also shows a comparison of transient response for an ideal interconnection (assuming direct connection) and Cu interconnection wire. Input signal is a 100 MHz square pulse. The propagation delays of MWCNT interconnects (β = 1 and β = 1/3) are close to ideal interconnect and are smaller than SWCNT bundle and Cu interconnects. As mentioned earlier, the number of SWCNTs in the bundle is larger than the number of shells in the MWCNT. Therefore, the resistance is smaller for the SWCNT bundle interconnect than that of the MWCNT. However, the capacitance of the SWCNT bundle interconnect is much larger than that of MWCNT. As a result, the propagation delay of MWCNT is smaller than that of the SWCNT bundle. The propagation delays are smaller for β = 1 than for β = 1/3 for both MWCNT and SWCNT bundle interconnects. This can be explained by the fact that there are more interconnect channels when β increases.

One of the advantages of CNT interconnect is its large mean free path, which is on the order of several microns as compared to ~40 nm for Cu at room temperature. It provides low resistivity and possible ballistic transport in short-length interconnects [45]. We have also simulated a CNT-FET inverter pair with 1 μm Cu, MWCNT and SWCNT bundle interconnects using Cadence/Spectre. Local interconnects are often used for connecting nearby gates or devices with lengths on the order of micrometers. Therefore, these have the smallest cross section and largest resistance per unit length compared to global interconnects. We have utilized the process parameters from the 2016 node for 22 nm technology [37] assuming 22 nm diameter of an MWCNT, 22 nm width and 44 nm thickness of an SWCNT bundle. Relatively global interconnects have larger cross section and smaller resistivity. The lengths are on the order of hundreds micrometers. We have utilized the process parameters from the 2016 node of 22 nm technology [37] assuming 33 nm diameter of an MWCNT, 33 nm width and 87 nm thickness of an SWCNT bundle. Simulations are performed for different lengths of Cu, MWCNT and SWCNT bundle interconnects corresponding to ballistic transport length (1 μm), local interconnects (10, 100 μm) to global interconnects (500 μm). The results are shown in Fig. 5.12. Dependence of delay on interconnection length in Fig. 5.12 shows that the increase in delay for Cu interconnects is larger than that of MWCNT and SWCNT bundle interconnects. The delays of MWCNT interconnects (β = 1 and β = 1/3) are smaller than that of SWCNT bundle and Cu interconnects. The delays are smaller for β = 1 than for $\beta = 1/3$ for both MWCNT and SWCNT bundle interconnects and is due to more interconnect channels with increase in β.

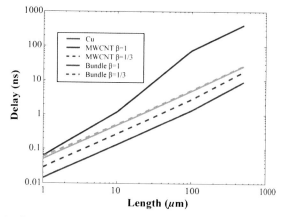

Figure 5.12 Propagation delays of interconnects of different lengths for 22 nm technology.

Power dissipation is another challenge to next-generation interconnects. We have simulated power dissipation for MWCNT and SWCNT bundle interconnects in 22 nm technology node and compared with the Cu wire interconnects. Table 5.3 summarizes power dissipation ratio of MWCNT and SWCNT bundle ($\beta = 1/3$ and $\beta = 1$) to Cu interconnect. CNT interconnects dissipates less power and especially for local interconnections. Maximum power dissipation in CNTs interconnections is no more than the 8% of the Cu interconnections.

Table 5.3 Power dissipation ratio of MWCNT and SWCNT bundle to Cu interconnects

	Normalized power dissipation (%)			
	Length (μm)			
Type of CNT	**1**	**10**	**100**	**500**
MWCNT ($\beta = 1$)	0.070	0.065	0.339	1.422
MWCNT ($\beta = 1/3$)	0.359	0.418	2.182	7.591
SWCNT Bundle ($\beta = 1$)	0.011	0.015	0.079	0.137
SWCNT Bundle($\beta = 1/3$)	0.036	0.047	0.256	0.688

Note: Normalization parameter is the length of Cu (1, 10, 100 and 500 μm). The technology node is 22 nm.

5.5   Summary

In this chapter, models for CNT interconnects, which include MWCNT and SWCNT bundle are discussed based on one-dimensional fluid theory. The one-dimensional fluid model can be applied to CNT interconnects using low-resistance contacts in current low-voltage nanometer CMOS technologies. The applicability of MWCNT and SWCNT bundle as interconnect wires for next-generation design of integrated circuits has been explored theoretically and compared with Cu interconnects in 22 nm technology node. Results of the one-dimensional fluid theory for SWCNT interconnect extended to MWCNT and SWCNT bundle interconnects show that MWCNT and SWCNT bundle interconnects have better performance than the Cu interconnects. MWCNT and SWCNT bundle interconnects exhibit higher transmission efficiency and lower reflection losses and less power dissipations. This is mainly due to larger conductivity of MWCNT and SWCNT bundle, proportional to the number of conducting shells (M) in MWCNT and conducting shells (N) in SWCNTs, respectively. With no special separation techniques, the metallic nanotubes are distributed with probability $\beta = 1/3$. While the proportion of metallic nanotubes can be potentially increased using techniques introduced by Peng et al. [31] and Zheng et al. [32], the delays in MWCNT and SWCNT bundle interconnects can be further decreased with increase in β and approaching to 1. It is also noticed that with the increase in interconnection length, the delay of Cu interconnect increases faster than that of MWCNT and SWCNT bundle interconnects. For applications that require small circuit delays, MWCNT interconnects should be used due to smaller capacitances. Applications requiring large transmission efficiency and low reflection losses, CNT bundles should be used for interconnects since the numbers of conducting channels per shell are more in SWCNTs bundle than the number of conducting channels per shell in MWCNT of the same size. These findings suggest that MWCNT and SWCNT bundle can replace Cu as interconnection wires in next generation of VLSI integrated circuits.

References

1. Raychowdhury, A., and Roy, K. (2004). Modeling and analysis of carbon nanotube interconnects and their effectiveness for high speed VLSI

design, *Proceedings of the 4th IEEE Conference on Nanotechnology* (2004). pp. 608–610.

2. Srivastava, N., and K. Banerjee, K. (2005). Performance analysis of carbon nanotube interconnects for VLSI applications, *Proceedings of the IEEE/ACM International Conference on Computer-Aided Design* (*ICCAD 2005*), pp. 383–390.

3. Salahuddin, S., Lundstrom, M., and Datta, S. (2005). Transport effects on signal propagation in quantum wires, *IEEE Trans. Electron Devices*, **52**, 1734–1742.

4. Nieuwoudt, A., and Massoud, Y. (2006). Understanding the impact of inductance in carbon nanotubes for VLSI interconnect using scalable modeling techniques, *IEEE Trans. Nanotechnol.*, **5**, 758–765.

5. Li, H., and Banerjee, K. (2009). High-frequency analysis of carbon nanotube interconnects and implications for on-chip inductor design, *IEEE Trans. Electron Devices*, **56**, 2202–2214.

6. Nieuwoudt, A., and Massoud, Y. (2006). Evaluating the impact of resistance in carbon nanotube bundles for VLSI interconnect using diameter-dependent modeling techniques, *IEEE Trans. Electron Devices*, **53**, 2460–2466.

7. Maffucci, A., Miano, G., and Villone, F. (2009). A new circuit model for carbon nanotube interconnects with diameter-dependent parameters, *IEEE Trans. Nanotechnol.*, **8**, 345–354.

8. Naeemi, A., and Meindl, J. D. (2006). Compact physical models for multiwall carbon-nanotube interconnect, *IEEE Electron Device Lett.*, **27**, 338–340.

9. Li, H., Yin, W.-Y., Banerjee, K., and Mao, J.-F. (2008). Circuit modeling and performance analysis of multi-walled carbon nanotube interconnects, *IEEE Trans. Electron Devices*, **55**, 1328–1337.

10. Fathi, D., Forouzandeh, B., Mohajerzadeh, S., and Sarvari, R. (2009). Accurate analysis of carbon nanotube interconnects using transmission line mode, *Micro Nano Lett.*, **4**, 116–121.

11. Burke, P. J. (2003). An RF circuit model for carbon nanotubes, *IEEE Trans. Nanotechnol.*, **2**, 55–58.

12. Burke, P. J. (2002). Lüttinger Liquid theory as a model of the gigahertz electrical properties of carbon nanotubes, *IEEE Trans. Nanotechnol.*, **1**, 129–144.

13. Maffucci, A., Miano, G., and Villone, F. (2008). A transmission line model for metallic carbon nanotube interconnects, *Int. J. Circuit Theory Appl.*, **36**, 31–51.

14. Nieuwoudt, A., and Massoud, Y. (2008). On the optimal design, performance, and reliability of future carbon nanotube-based interconnect solutions, *IEEE Trans. Electron Devices*, **55**, 2097–2110.

15. Nihei, M., Kondo, D., Kawabata, A., Sato, S., Shioya, H., Sakaue, M., Iwai, T., Ohfuti, M., and Awano, Y. (2005). Low-resistance multi-walled carbon nanotube vias with parallel channel conduction of inner shells, *Proceedings of the IEEE International Interconnect Technology Conference*, pp. 234–236.

16. Li, H. J., Lu, W. G., Li, J. J., Bai, X. D., and Gu, C. Z. (2005). Multichannel ballistic transport in multiwall carbon nanotubes, *Phys. Rev. Lett.*, **95**, 86601-1 to 86601-4.

17. Yan, Q., Wu, J., Zhou, G., Duan, W., and Gu, B.-L. (2005). Ab initio study of transport properties of multiwalled carbon nanotubes, *Phys. Rev. B*, **72**, 155425-1 to 155425-5.

18. Massoud, Y., and Nieuwoudt, A. (2006). Modeling and design challenges and solutions for carbon nanotube-based interconnect in future high performance integrated circuits, *ACM J. Emerg. Technol. Comput. Syst.*, **2**, 155–196.

19. Srivastava, A., Xu, Y., and Sharma, A. K. (2010). Carbon nanotubes for next generation very large scale integration interconnects, *J. Nanophoton.*, **4**, 1–26.

20. Forró, L., and Schönenberger, C. (2000). *Carbon Nanotubes: Synthesis, Structure, Properties and Applications*, (Springer-Verlag Berlin, Germany).

21. Datta, S. (2005). *Quantum Transport: Atom to Transistor*, (Cambridge University Press, London, UK).

22. Xu, Y., and Srivastava, A. (2010). A model for carbon nanotube interconnects, *Int. J. Circ. Theor. Appl.*, **38**, 559–575.

23. Rosenthal, D. M., and Asimow, R. M. (1971). *Introduction to Properties of Materials*, (Van Nostrand Reinhold, New York).

24. Miano, G., and Villone, F. (2006). An integral formulation for the electrodynamics of metallic carbon nanotubes based on a fluid model, *IEEE Trans. Antennas Propagation*, **54**, 2713–2724.

25. Chiariello, A. G., Maffucci, A., Miano, G., Villone, F., and Zamboni, W. (2006). Metallic carbon nanotube interconnects, part I: A fluid model and a 3D integral formulation, *Proceedings of the IEEE Workshop on Signal Propagation on Interconnects*, pp. 181–184.

26. Ramo, S., Whinnery, J. R., and Duzer, T. V. (1994). *Fields and Waves in Communication Electronics*, (Wiley, New York).

27. Xu, Y., Srivastava, A., and Sharma, A. K. (2009). A model of multi-walled carbon nanotube interconnects, *Proceedings of the 52nd IEEE Midwest Symposium on Circuits and Systems*, pp. 987–990.

28. Thess, A., Lee, R., Nikolaev, P., Dai, H., Petit, P., Robert, J., Xu, C., Lee, Y. H., Kim, S. G., Rinzler, A. G., Colbert, D. T., Scuseria, G. E., Tománek, D., Fischer, J. E., and Smalley, R. E. (1996). Crystalline ropes of metallic carbon nanotubes, *Science*, **273**, 483–487.

29. Avouris, P., Appenzeller, J., Martel, R., and Wind, S. J. (2003). Carbon nanotube electronics, *Proc. IEEE*, **91**, 1772–1783.

30. Liebau, M., Graham, A. P., Duesberg, G. S., Unger, E., Seidel, R., and Kreupl, F. (2005). *Fullerenes, Nanotubes, Carbon Nanostruct.*, **13**, 255258.

31. Peng, N., Zhang, Q., Li, J., and Liu, N. (2006). Influences of ac electric field on the spatial distribution of carbon nanotubes formed between electrodes, *J. Appl. Phys.*, **100**, 024309-1 to 024309-5.

32. Zheng, M., Jagota, A., Strano, M. S., Santos, A. P., Barone, P., Chou, S. G., Diner, B. A., Dresselhaus, M. S., McLean, R. S., Onoa, G. B., Samsonidze, G. G., Semke, E. D., Usrey, M., and Walls, D. J. (2003). Structure-based carbon nanotube sorting by sequence-dependent DNA assembly, *Science*, **302**, 1545–1548.

33. Xu, Y., Srivastava, A., and Sharma, A. K. (2010). Emerging carbon nanotube electronic circuits, modeling and performance, *VLSI Des.*, **2010**, 1–8.

34. Naeemi, A., Davis, J. A., and Meindl, J. D. (2004). Compact physical models for multilevel interconnect crosstalk in gigascale integration (GSI), *IEEE Trans. Electron Devices*, **51**, 1902–1912.

35. Naeemi, A., and Meindl, J. D. (2007). Design and performance modeling for single-walled carbon nanotubes as local, semiglobal, and global interconnects in gigascale integrated systems, *IEEE Trans. Electron Devices*, **54**, 26–37.

36. Slepyan, G. Y., Krapivin, N. A., Maksimenko, S. A., Lakhtakia, A., and Yevtushenko, O. M. (2001). Scattering of electromagnetic waves by a semi-infinite carbon nanotube, *Intl. J. Electron. Commun.*, **55**, 273–280.

37. *International Technology Roadmap for Semiconductors*. Available: http://www.itrs.net/Links/2007ITRS/Home2007.htm(2007).

38. Close G. F., and Wong, H. S. P. (2007). Fabrication and characterization of carbon nanotube interconnects, *IEDM Tech. Dig.*, 203–206.

39. Close, G. F., Yasuda, S., Paul, B., Fujita, S., and Wong, H.-S. P. (2008). A 1 GHz integrated circuit with carbon nanotube interconnects and silicon transistors, *Nano Lett.*, **8**, 706–709.

40. Close, G. F., Yasuda, S., Paul, B. C., Fujita, S., and Wong, H.-S. P. (2009). Measurment of subnanosecond delay through multiwall carbon-nanotube local interconnects in a CMOS integrated circuit, *IEEE Trans. Electron Devices*, **56**, 43–49.

41. Slepyan, G. Y., Maksimenko, S. A., Lakhtakia, A., Yevtushenko, O., and Gusakov, A. V. (1999). Electrodynamics of carbon nanotubes: dynamic conductivity, impedance boundary conditions, and surface wave propagation, *Phys. Rev. B*, **60**, 17136.

42. Srivastava, N., Li, H., Kreupl, F., and Banerjee, K. (2009). On the applicability of single-walled carbon nanotubes as VLSI interconnects, *IEEE Trans. Nanotechnol.*, **8**, 542–559.

43. Srivastava, A., Marulanda, J. M., Xu, and Y., and Sharma, A. K. (2009). Current transport modeling of carbon nanotube field effect transistor, *Phys. Status Solidi* (*A*), **206**, 1569–1578.

44. Xu, Y., and Srivastava, A. (2009). Dynamic response of carbon nanotube field effect transistor circuits, *Proceedings of the 2009 NSTI Nanotechnology Conference and Expo*, pp. 625–628.

45. McEuen, P. L., Fuhrer, M. S., and Park, H. (2002). Single-walled carbon nanotube electronics, *IEEE Trans. Nanotechnol.*, **1**, 78–85.

Chapter 6

Carbon Nanotube Wire Inductors

6.1   Introduction

It has been demonstrated that the carbon nanotube (CNT) wire is very likely to replace the Cu interconnect in sub-nanometer CMOS technologies [1]. It has also been shown that carbon nanotube wire has reduced skin effect compared to metal conductors such as the Cu and has a great promise for realization of high-Q on-chip inductors for RF integrated circuits [2]. Several models of CNT based on-chip inductor have been presented in the literature [2–5], and a method of fabrication of a planar spiral CNT inductor has also been proposed [2]. Recently, Srivastava [6] has proposed the use of high-Q on-chip CNT wire inductors in design of LC-VCO (voltage-controlled oscillators) for phase-locked loop (PLL) systems.

Phase-locked loops are widely used in high-speed and high-frequency data communication systems. One of the important building blocks of the PLL is voltage-controlled oscillator (VCO). Digital cellular communication devices operating in GHz range widely employ VCOs. Commonly used VCO employ LC tuned circuit where quality factor of the inductor becomes crucial to the operation of the oscillator. In the past, inductor has been realized from bonding wires to retain large quality factor [7]. With shrinking device geometries and packaging requirement on-chip inductors

Carbon-Based Electronics: Transistors and Interconnects at the Nanoscale
Ashok Srivastava, Jose Mauricio Marulanda, Yao Xu, and Ashwani K. Sharma
Copyright © 2015 Pan Stanford Publishing Pte. Ltd.
ISBN 978-981-4613-10-1 (Hardcover), 978-981-4613-11-8 (eBook)
www.panstanford.com

have been realized for radio frequency integrated circuits using Al and Cu metallization. Salimath [8] has reviewed the design of several CMOS voltage controlled oscillators and presented an on-chip 1.1 to 1.8 GHz VCO implementation in CMOS for use in RF integrated circuits. However, achieving high-Q inductor is still being researched.

Phase-locked loops operating in 1–2 GHz range suffer from phase noise which results in degradation in performance of RF systems where high frequency phase-locked loops are used. The LC VCO in PLL is the key contributor to the phase noise, which uses metallic wire on-chip integrated inductor. The resistive losses lower the inductor Q factor in such LC VCO design. In this chapter, we present design of a new 2 GHz LC VCO in TSMC 0.18 µm CMOS process using multi-walled carbon nanotube (MWCNT) and single-walled carbon nanotube(SWCNT) bundle wires as an inductor in the LC tank circuit. We have applied our CNT interconnect model, which is described in our previous work [9], in a well-known π model [10] to study the properties of MWCNT and SWCNT bundle wire on-chip inductors. The structure of an SWCNT can be conceptualized by wrapping an atomic thick layer of graphite called graphene into a seamless cylinder. Multi-walled carbon nanotubes consist of multiple layers of graphite rolled in to form a tubular shape. We have calculated Q factors for MWCNT and SWCNT bundle wire inductors and Cu wire inductor for comparison and studied performance of LC VCO.

6.2 On-Chip Inductor Modeling

The widely used π model [10] is utilized to model the on-chip inductor as shown in Fig. 6.1. The model elements of Fig. 6.1 are described as follows. The series inductance and resistance of interconnects are described by L_S and R_S, respectively. The capacitance, C_S models the feed-through path between two terminals of the inductor. The capacitance, C_{OX} models the oxide capacitance between spiral inductor and the silicon substrate. The capacitance, C_{sub} and the resistance, R_{sub} model the capacitance and the resistance of the silicon substrate, respectively. The π model was originally proposed for integrating on-chip spiral inductors using metal layers such as the aluminum and copper in a silicon substrate. The π model also includes the eddy currents at high frequencies described by the resistance, R_{eddy} in series with the inductance, L_S and resistance, R_S. Since LC VCO is designed using on-chip spiral inductors from metallic MWCNT and bundles of SWCNTs, the π model proposed by Yue and Wong [10] is used in the present work.

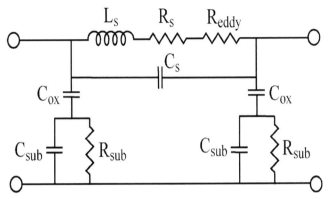

Figure 6.1 π Model of an on-chip inductor.

The ac electrical resistivity of copper can be predicted by Drude model [11]:

$$\rho(\omega) = \rho_0(1 + j\omega\tau), \qquad (6.1)$$

where ρ_0 is dc resistivity and τ is the momentum relaxation time.

The relation between ac electrical conductivity of CNT and frequency is given by [12],

$$\sigma = \frac{\sigma_0}{1 + i\omega\tau}, \qquad (6.2)$$

where $\sigma_0 = 1/R_S$, $\tau = l_{mfp}/v_F$ is electron relaxation time in CNT, v_F is the Fermi velocity, and l_{mfp} is mean-free path.

At high frequencies, the Eddy currents induced in the substrate will significantly decrease the performance of the inductor. Approximate expression for Reddy is given by [3],

$$R_{eddy} = \frac{8l}{\sigma t W}, \qquad (6.3)$$

where l, t, and W are length, thickness and width of the interconnect, respectively, and σ is the conductivity of the material.

In Fig. 6.1, capacitance (C_S), oxide layer capacitance (C_{ox}), substrate resistance (R_{sub}) and substrate capacitance (C_{sub}) are calculated by using the modeling techniques presented in [10] based on the total length of the inductor. Table 6.1 summarizes the calculated values of π model parameters of Fig. 6.1, which are used in simulations. In Table 6.1, L_S = 6 nH is the designed value for the LC VCO for both the metallic carbon nanotubes (MWCNT and SWCNT bundle) and copper wire inductors. The model parameters except R_S are same for both metallic carbon nanotubes and copper because these are dependent on the geometry of the spiral inductor and independent on the metal materials. Here, β is the probability factor characterizing an SWCNT being metallic in a bundle or a shell being metallic in an MWCNT.

Table 6.1   Inductor π model parameters

Parameter	MWCNT (β = 1/3)	MWCNT (β = 1)	SWCNT bundle (β = 1/3)	SWCNT bundle (β = 1)	Cu
R_S (Ω)	1.8	0.6	0.18	0.06	5
R_{eddy} (Ω)	0.01	0.01	0.01	0.01	0.01
C_{OX} (fF)	4.7	4.7	4.7	4.7	4.7
C_S (aF)	38	38	38	38	38
R_{sub} (MΩ)	5	5	5	5	5
C_{sub} (fF)	0.1	0.1	0.1	0.1	0.1
L_S (nH)	6	6	6	6	6

The performance of CNT bundle wire and MWCNT wire inductors are analyzed and compared to that of Cu inductors. We studied the utilization of CNT bundle and MWCNT wire inductors in 0.18 μm CMOS technology. The inductors considered is a 4.5 turn planar spiral inductor, which has the outermost diameter D_{out} = 250 μm, conductor width W = 15 μm, conductor thickness t = 2 μm, conductor spacing S = 1.5 μm, oxide and substrate thicknesses of 4 nm and 300 μm, respectively.

The Q factor analysis results are shown in Fig. 6.2. The maximum Q factor of SWCNT bundle (β = 1) inductor can be ~600% higher than that of the Cu inductor and the maximum Q factor of MWCNT (β = 1) inductor can be ~200% higher than that of the Cu inductor. This significant enhancement in Q factor arises not only because of the lower resistance of CNTs but also because the skin effect in CNT interconnects is negligible [13]. The Q factors of CNT bundle wire inductors are much higher than that of the MWCNT wire inductor because there are more conductance channels in a bundle compared with a same size MWCNT wire, which means the resistance of an SWCNT bundle wire is smaller than that of a same size MWCNT wire.

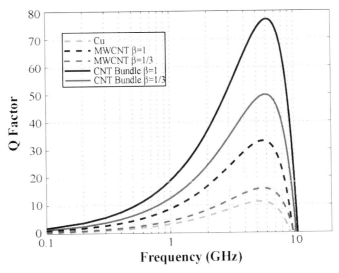

Figure 6.2 Q factor of inductors based on CNT and Cu.

6.3   LC Voltage-Controlled Oscillator

The primary goal in the design of the oscillator is to design active devices to overcome the losses associated with the tank parallel resistance. A cross-coupled CMOS differential oscillator shown in Fig. 6.3a was chosen due to its better phase noise performance compared to the cross-coupled single type transistor (NMOS or PMOS) topologies [8]. It consists of three components: LC tank, tail bias transistor and cross-coupled differential pair. LC tank is made by an inductor and a capacitor connected in cascade or in parallel. Figure 6.3b shows the equivalent circuit of this LC VCO. Here R_C and R_L are the resistance of capacitor and inductor, respectively, and R_P is frequency-dependent shunt resistance.

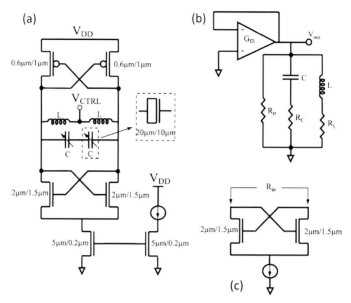

Figure 6.3 (a) Circuit diagram of a CMOS cross-coupled LC VCO, (b) equivalent circuit of LC VCO, and (c) circuit to estimate the negative resistance of the cross-coupled pair.

To compute the G_m of the amplifier, we need first to consider the cross-coupled NMOS transistors shown in Fig. 6.3c. The cross-coupled CMOS transistors form the negative resistance. This negative resistance is used to offset the positive resistance in the passive components, L and C to produce an oscillation [14]. The resistance, R_{in} seen at the drain of the NMOS transistor pair is given by

$$R_{in} = -\frac{2}{g_m}, \tag{6.4}$$

where g_m is the small-signal transconductance of each transistor. Therefore, the total transconductance of the CMOS pair can be expressed as a parallel combination of the NMOS and PMOS transistor pair negative resistance, R_{inn} and R_{inp}:

$$G_m = \frac{1}{R_{inn} // R_{inp}}, \tag{6.5}$$

The resonance frequency of LC tank circuit is given by

$$f_0 = \frac{1}{2\pi\sqrt{LC}} , \tag{6.6}$$

We designed a 2 GHz LC VCO in TSMC 0.18 μm CMOS process as shown in Fig. 6.3a where an inductor in the LC tank circuit is realized from a CNT bundle wire and MWCNT wire. The symmetrical design of the VCO gives good phase noise performance and large voltage swing. The varactor in the circuit of Fig. 6.3a is implemented from an nMOSFET with source and drain tied together. The C–V curve of the varactor is shown in Fig. 6.4. The voltage-controlled capacitance range is from 200 fF to 2.2 pF, which makes the LC VCO to oscillate from 1.6 GHz to 3.3 GHz. If we choose the capacitor's value to be 0.5 pF, from Eq. (6.6), to get the LC VCO oscillation at 3 GHz, the value of the inductor is calculated to be 6 nH. In our 2 GHz LC VCO design, we have chosen 6 nH inductor since effective value of inductor may be less due to associated parasitics.

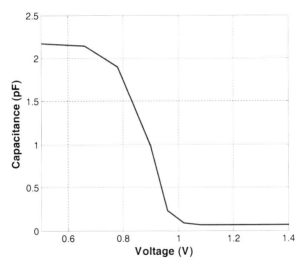

Figure 6.4 C–V curve of a varactor.

Figure 6.5 shows the LC VCO oscillation frequency versus control voltage for inductor with no losses (ideal), SWCNT bundle (β = 1, 1/3) and MWCNT (β = 1, 1/3) wire inductors and the Cu wire inductor. It is also shown that LC VCO with MWCNT wire inductors have higher oscillation frequencies than that with the Cu wire inductor. CMOS LC VCO with SWCNT bundle (β = 1/3) wire inductor oscillates between 1.3 and 2.6 GHz, which is higher in comparison to oscillation frequency range 1.1–2.4 GHz of LC VCO with Cu wire inductor. It is also shown that VCO with SWCNT bundle wire inductor (β = 1) has higher oscillation frequency than that with the Cu wire inductor. For example, oscillation frequency of CNT bundle (β = 1) wire inductor is higher by ~10% at 1.2 V than that with Cu wire inductor. Moreover at 1.2 V control voltage, the Q factor of Cu inductor is 7 and Q factors of SWCNT bundle inductors are 38 (for β = 1) and 26 (for β = 1/3). Also, the oscillating frequency is higher in VCO with MWCNT and SWCNT bundle wire inductors for β = 1 than that with β = 1/3.

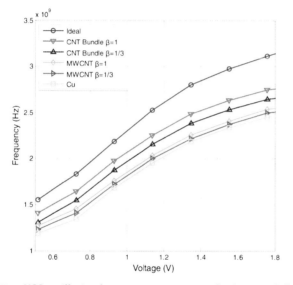

Figure 6.5   VCO oscillation frequency versus control voltage with difference inductors.

The phase noise is modeled by using the Leeson's phase noise density equation [15]:

$$L(\Delta f) = \left(\frac{1}{8Q^2} \right) \left(\frac{FkT}{P} \right) \left(\frac{f_0}{\Delta f} \right)^2, \tag{6.7}$$

where k is Boltzann's constant, T is the temperature, P is the output power, F is the noise factor, Q is the quality factor of the LC tank, f_0 is the oscillation frequency, and Δf is the offset frequency from f_0.

Since resistance of MWCNT and SWCNT bundle wire inductors are smaller than that of Cu wire inductor, the losses are smaller, which make the Q of LC tank circuit higher. Therefore, phase noise decreases following Eq. (6.7). Figure 6.6 shows the phase noise of the LC VCO at 2 GHz tuning frequency. The LC VCO phase noise using inductor with no losses (ideal) is –70 dBc/Hz at 10 kHz offset frequency from the carrier and –123 dBc/Hz at 10 MHz offset frequency from the carrier.

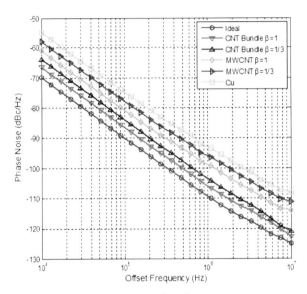

Figure 6.6   VCO phase noise versus offset frequency.

The LC VCO with SWCNT bundle wire inductors has about 10 dBc/Hz smaller phase noise than that with Cu wire inductors. Moreover, at this 2 GHz tuning frequency, the Q factor of Cu inductor is 7 and Q factor of CNT bundle inductors are 36 (for $\beta = 1$) and 25 (for $\beta = 1/3$). The phase noise is smaller in LC VCO with SWCNT bundle wire inductors for $\beta = 1$ than $\beta = 1/3$. The VCO with MWCNT wire inductors has about 5 dBc/Hz smaller phase noise than that with Cu wire inductors. The phase noise is smaller in LC VCO with MWCNT inductors for $\beta = 1$ than $\beta = 1/3$.

When Q factors of Cu, MWCNT and SWCNT bundle wire inductors are compared, the SWCNT bundle wire inductor shows high Q at the resonance. Its effect is clearly reflected on the oscillation frequency and phase noise of LC VCO simulations. The

LC VCO shows significantly improved performance in oscillation frequency and phase noise using CNT wire inductors as compared to the Cu wire inductor. Furthermore, LC VCO using SWCNT bundle wire inductor shows better performance that the LC VCO using MWCNT wire inductor.

6.4 Summary

We have utilized MWCNT and SWCNT bundle interconnects model in a widely used π model to study the performances of MWCNT and SWCNT bundle wire inductors and compared them with Cu inductors [16]. The calculation results show that the Q factors of CNT wire (bundle and MWCNT) inductors are higher than that of the Cu wire inductor. This is mainly due to much lower resistance of CNT and negligible skin effect in carbon nanotubes at higher frequencies. The application of CNT wire inductor in LC VCO is also studied and the Cadence/Spectre simulations show that VCOs with CNT bundle wire inductors have significantly improved performance such as the higher oscillation frequency, lower phase noise, due to their smaller resistances and higher Q factors. It is also noticed that CMOS LC VCO using SWCNT bundle wire inductor has better performance when compared with the performance of LC VCO using MWCNT wire inductor due to its lower resistance and higher Q factor.

References

1. Li, H., Xu, C., Srivastava N., and Banerjee, K. (2009). Carbon nano-materials for next-generation interconnects and passives: Physics, status and prospects, *IEEE Trans. Electron Devices*, **56**, 1799–1821.

2. Li, H., and Banerjee, K. (2009). High-frequency analysis of carbon nanotube interconnects and implications for on-chip inductor design, *IEEE Trans. Electron Devices*, **56**, 2202–2214.

3. Kuhn, W. B., and Ibrahim, N. M. (2001). Analysis of current crowding effects in multiturn spiral inductors, *IEEE Trans. Microwave Theory Tech.*, **49**, 31–38.

4. Nieuwoudt, A., and Massoud, Y. (2006). Understanding the impact of inductance in carbon nanotube bundles for VLSI interconnect using scalable modeling techniques, *IEEE Trans. Nanotechnol.*, **5**, 758–765.

5. Nieuwoudt, A., and Massoud, Y. (2008). Predicting the performance of low-loss on-chip inductors realized using carbon nanotube bundles, *IEEE Trans. Electron Devices*, **55**, 298–312.

6. Srivastava, A. (2009). *Carbon Nanotube (CNT) Wire Inductor for CMOS LC VCO in PLL Synthesizer*, Tech. Disclosure, OIP (Louisiana State University, U.S.A.).

7. Hung, C. M., and O, K. K. (2000). A packaged 1.1-GHz CMOS VCO with phase noise of –126 dBc/Hz at a 600-kHz offset, *IEEE J. Solid-State Circuits*, **35**, 100–103.

8. Salimath, C. S. (2006). *Design of CMOS LC Voltage Controlled Oscillators*, M. S. (EE) Thesis (Louisiana State University, U.S.A.).

9. Srivastava, A., Xu, Y., and Sharma, A. K. Carbon nanotubes for next generation very large scale integration interconnects. *J. Nanophoton.*, **4**, 1–26.

10. Yue, C. P., and Wong, S. S. (2000). Physical modeling of spiral inductors on silicon, *IEEE Trans. Electron Devices*, **47**, 560–568.

11. Ashcroft, N. W., and Mermin, N. D. (1976). *Solid State Physics* (Saunders College, Philadelphia, PA).

12. Slepyan, G. Y., Maksimenko, S. A., Lakhtakia, A., Yevtushenko O., and Gusakov, A. V. (2009). Electrodynamics of carbon nanotubes: Dynamic conductivity, impedance boundary conditions, and surface wave propagation, *Phys. Rev. B*, **60**, 17136–17149.

13. Banerjee, K., Li, H., and Srivastava, N. (2008). *Proceedings of the 8th IEEE Conference on Nanotechnology (NANO)*, pp. 432–436.

14. Lee, T. H. (2004). *The Design of CMOS Radio-Frequency Integrated Circuits* (Cambridge University Press, New York).

15. Leeson, D. B. (1966). A simple model of feedback oscillator noise spectrum. *Proc. IEEE*, **54**, 329–330.

16. Srivastava, A., Xu, Y., Liu, Y. Sharma, A. K., and Mayberry, C. (2012). CMOS LC voltage controlled oscillator design using carbon nanotube wire inductors, *ACM J. Emerg. Technol. Comput. Syst.* (Special Issue), 8(3), Article **15**, 15.1–15.9.

Chapter 7

Energy Recovery Techniques for CNT-FET Circuits

7.1   Introduction

Energy recovery techniques are playing an important role in modern electronic circuit design due to urgent need of low power dissipation. More so with increasing demand for portable communication and computing systems, power dissipation has become one of the major chip design concerns. Circuit-level techniques, including energy recovery techniques, have been researched for low-power design. Compared with energy recovery techniques, special cooling techniques such as cryogenics [1] would demand a paradigm shift in the technology and would not be cost-effective. Aggressive scaling of the silicon technology has always been associated with the need to realize low power at higher performance. The conventional approaches for active power reduction are to scale down the supply voltage, to decrease the load capacitances and to reduce signal transitions. However, to maintain sufficient noise margin, a reasonable supply voltage is required. The device dimension and the associated parasitic limit load capacitance scaling. With increasing processing speed, the number of signal transitions per unit time is fast increasing. Furthermore, an exponential increase in transistor count per chip has led to an alarming increase in the power density on the chip. Advancing

y

Carbon-Based Electronics: Transistors and Interconnects at the Nanoscale
Ashok Srivastava, Jose Mauricio Marulanda, Yao Xu, and Ashwani K. Sharma
Copyright © 2015 Pan Stanford Publishing Pte. Ltd.
ISBN 978-981-4613-10-1 (Hardcover), 978-981-4613-11-8 (eBook)
www.panstanford.com

into an era of nanotechnologies, device dimensions would shrink further and on-chip power density would reach such magnitudes that conventional cooling techniques would not be able to handle. Innovations in the cooling technology should thus be augmented with new ways of circuit design such that the heat dissipated on a chip can be lowered and kept within certain limits.

Currently carbon nanotube field-effect transistors (CNT-FETs) are one of the promising candidates for next generation transistors in future electronic circuit design. With its highly scaled dimensions and high current density, CNT-FET would increase the power density on a chip. It is predicted that it would far exceed the maximum power density limitation [2]. Therefore, low-power designs are necessary for the design of CNT-FET integrated circuits, especially in the high-frequency operation region.

Energy loss or heat dissipation in a circuit is caused by the following three mechanisms:

(1) Energy loss occurs when there is a voltage difference between the source and drain terminals and the transistor turns on. Energy loss in a conventional logic belongs to this category.

(2) Energy loss due to charge transfer from one capacitor to other, when these are connected by a switch of finite resistance.

(3) Subthreshold operation and gate tunneling cause an increase in leakage current and results in energy loss.

Thus, on-chip power densities will be a big challenge, and it is where the energy recovery techniques are expected to play an important role in electronic circuit designs. Energy recovery techniques for CMOS circuits are well developed, and these techniques can be applied to CNT-FET integrated circuit design to examine the effect of power density. Hwang et al. [3] have applied CMOS energy recovery techniques in CNT-FET circuits. However, more research is needed to develop an energy efficient recovery technique for CNT-FET circuits. The exclusive-OR (XOR) and exclusive-NOR (XNOR) gates are well known for their roles in larger circuits such as full adders and parity checkers [4]. Therefore, an optimized design of XNOR/XOR gates using energy recovery techniques can certainly benefit the performance of the larger circuits where these are used. In this chapter, we present the design of energy recovery CNT-FET circuits that is based on our earlier work in energy recovery CMOS circuits and CNT-FET models [5–7].

7.2   CNT-FET Models

Current transport equations for CNT-FETs shown in Fig. 7.1 have been discussed in detail in our recent work [6,7]. In addition, we have used a Meyer capacitance model [7–9] to characterize dynamic response of CNT-FETs [7].

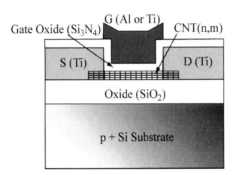

Figure 7.1   Plot of the vertical cross section of a CNT-FET.

CNT-FETs do not quite behave the same way as the traditional CMOS-FETs and equations are to be obtained to describe the current transport which may not be compatible with SPICE. In such situation, Verilog-Analog Mixed-Signal (Verilog-AMS) can be used for simulation. Figure 7.2 shows the steps needed in using Verilog-AMS in Cadence® Spectre.

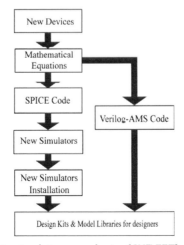

Figure 7.2   Steps to simulate a new device (CNT-FET) using Verilog-AMS.

7.3 Energy Recovery Logic (ERL) and Circuit Simulation Results

Different energy recovery logic (ERL) families have been proposed in [3,10–13] for CMOS circuits and have been used in this work to estimate reduction in on-chip power density of CNT-FET circuits. We reviewed these ERL devices in our previous work [5] and compared their performances in CMOS circuits. In this work, we studied the application of quasi-static energy recovery logic (QSERL) [12], 2N2N2P [10], clocked adiabatic logic (CAL) [13], and new CAL [5] technologies in CNT-FET circuits to show their performances in reducing the power density.

Figure 7.3 shows new improved CMOS CAL XNOR/XOR, which we have proposed in our previous work [5]. Cross-coupled inverters provide the memory function. Auxiliary timing control clock signal CX is used to realize an adiabatic inverter and logic functions with a single power clock [12]. This signal controls transistors M1 and M2, which are in series with the input logic trees. We have modified CMOS trees to replace traditional NMOS logic trees [14] so that these do not need additional power supply since separated by M3 and M4.

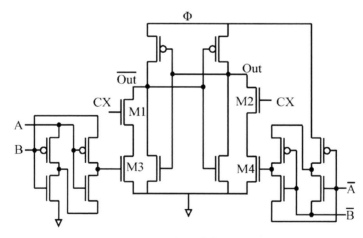

Figure 7.3   CMOS new CAL XNOR/XOR [5].

Input timing waveforms that we have used in new CAL XNOR/XOR are the same as in traditional CAL and shown in Fig. 7.4; Φ is trapeziform wave as a power clock and CX is an auxiliary clock to enable the gate function. The inputs are two complementary trapeziform waves. In the evaluation phase, the auxiliary clock CX enables the logic evaluation.

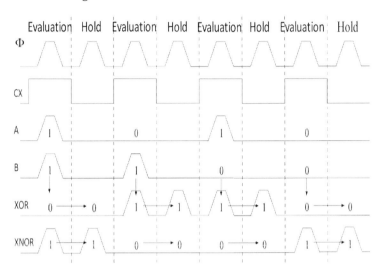

Figure 7.4   Ideal new CAL input waveforms.

The CMOS new CAL XNOR/XOR can be easily implemented using CNT-FETs. Figure 7.5 shows the new CAL XNOR/XOR using complementary CNT-FETs.

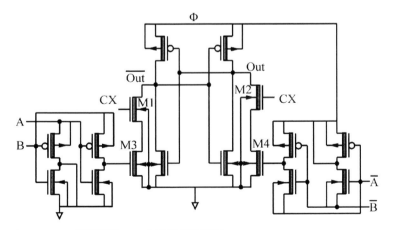

Figure 7.5   CNT-FET new CAL XNOR/XOR.

The power density on a chip, P_D can be calculated as follows [3]:

$$P_D = \frac{P}{\alpha A},\qquad(7.1)$$

where P is the power consumed. A is the total transistor active area and α is the area factor that accounts for the chip area due to routing. Figure 7.6 shows the average power density of a CNT-FET inverter. We have assumed $\alpha = 4$ [3] in our simulations. Figure 7.6 shows that the power density of a CNT-FET inverter is higher than the limit of on-chip power density predicted by ITRS 2003 [2], especially at high-frequency operation range. Therefore, to design CNT-FET integrated circuits, at high-frequency operation range, low-power designs, such as energy recovery techniques, are necessary to keep the power density lower than the limits predicted by ITRS 2003 [2].

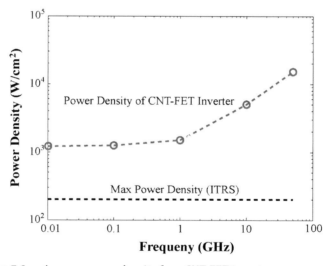

Figure 7.6   Average power density for a CNT-FET inverter.

Figure 7.7 shows the simulation results of power densities of different CNT-FET energy recovery XNOR/XOR circuits at different frequencies. The results in Fig. 7.7 show that new CAL ERL has the least power density. However, the power densities of energy recovery circuits described exceed the maximum power density limit over 1 GHz.

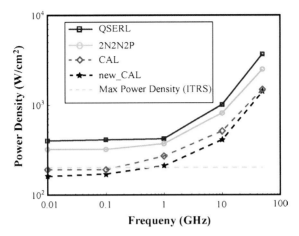

Figure 7.7   Average power density of different CNT-FET XNOR/XOR energy recovery circuits.

7.4   Summary

It is important to adopt circuit design techniques that would reduce on-chip power density for the future technology generations using CNT-FETs. In this work, using our CNT-FET models, we have studied the power density of CNT-FET circuits, which exceeds the maximum power density limit set by the ITRS 2003. Our simulations show that energy recovery techniques help in reducing the power density of CNT-FET circuits below 1 GHz. Beyond 1 GHz, further work is needed to reduce the on-chip power density. Energy recovery, thus, can be used as an alternative approach toward circuit designs for reduction of high power density.

References

1. Hutchby, J. A., Bourianoff, G. I., Zhirnov, V. V., and J. E. Brewer. (2002). Extending the road beyond CMOS, *IEEE Circuits Devices Mag.*, **18**, 28–41.

2. International Technology Roadmap for Semiconductors (http://www.itrs.net/Links/2003ITRS/Home2003.htm).

3. Hwang, M. E., Raychowdhury, A., and Roy, K. (2005). Energy-recovery techniques to reduce on-chip power density in molecular nanotechnologies, *IEEE Trans. Circuits Syst. I: Regular Papers*, **52**, 1580–1589.

4. Wang, J.-M., Fang, S.-C., and Feng, W.-S. (1994). New efficient designs for XOR and XNOR functions on the transistor level, *IEEE J. Solid-State Circuits*, **29**, 780–786.

5. Xu, Y., and Srivastava, A. (2007). New energy recovery CMOS XNOR/ XOR gates, *IEEE Proceedings 50th Midwest Symposium on Circuits and Systems* (*MWSCAS*), pp. 948–951.

6. Srivastava, A., Marulanda, J. M., Xu, Y., and Sharma, A. K. (2009). Current transports modeling of carbon nanotube fields effect transistors, *Phys. Status Solidi (A)*, **206**, 1569–1578.

7. Xu, Y., and Srivastava, A. (2009). Transient behavior of integrated carbon nanotube field-effect transistor circuits and bio-sensing applications, *Proc. SPIE: Nano. Bio. Info-Tech Sensors Syst.*, **7291**, 72910I-1 to 72910I-11.

8. Fjeldly, T. A., Ytterdal, T., and Shur, M. S. (1998). *Introduction to Device Modeling and Circuit Simulation* (Wiley, New York).

9. Cheng, Y., and Hu, C. (1999). *MOSFET Modeling and BSIM3 User's Guide* (Springer).

10. Kramer, A., Denker, J. S., Flower, B., and Moroney, J. (1995). 2nd order adiabatic computation with 2N-2P and 2N-2N2P logic circuits, *Proc. Intl. Symp. Low Power Des.*, 191–196.

11. Yong, M., and Deog-Kyoon, J. (1996). An efficient charge recovery logic circuit, *IEEE J. Solid-State Circuits*, **31**, 514–522.

12. Ye, Y., and Roy, K. (2001). QSERL: Quasi-static energy recovery logic, *IEEE J. Solid-State Circuits*, **36**, 239–248.

13. Maksimovic D., and Oklobdžija, V. G. (1995). Clocked CMOS adiabatic logic with single AC power supply, *Proc. 21st European Solid State Conference*, pp. 370–373.

14. Maksimovic, D., Oklobdzija, V. G., Nikolic, B., and Current, K. W. (2000). Clocked CMOS adiabatic logic with integrated single-phase power-clock supply, *IEEE Trans. on Very Large Scale Integration (VLSI) Systems*, **8**, pp. 460–463.

Chapter 8

Verilog-AMS Codes for Non-Ballistic CNT-FET Modeling

8.1   Introduction

As discussed in Chapter 3, new carbon nanotube field-effect transistors (CNT-FETs) do not behave the same way as the traditional CMOS-FETs. New model equations are to be developed to describe the current, capacitance, and so forth. Therefore, popular IC design and analysis simulator SPICE is not keeping up with the development of new device models but Verilog Analog Mixed-Signal (Verilog-AMS) gives a solution. Figure 7.2 shows that much less steps are needed to get the simulation available using Verilog-AMS compared with the SPICE coding. This is because model equations for new devices can be put into the Verilog-AMS coding and simulator will call the code. On the other hand, the model equations need to be fit into the SPICE simulator, which means a new simulator need to be built and this is a time consuming job. In our work, we have used Verilog-AMS to describe the CNT-FET static and dynamic models.

8.2   Verilog-AMS Code for n-Type CNT-FET

We have implemented our CNT-FET model in Verilog-AMS code and used it in our study of CNT-FET circuits. Following is the code

Carbon-Based Electronics: Transistors and Interconnects at the Nanoscale
Ashok Srivastava, Jose Mauricio Marulanda, Yao Xu, and Ashwani K. Sharma
Copyright © 2015 Pan Stanford Publishing Pte. Ltd.
ISBN 978-981-4613-10-1 (Hardcover), 978-981-4613-11-8 (eBook)
www.panstanford.com

for n-type CNT-FET. Code for p-type is similar, by changing the current direction.

```
// VerilogA for CNTFET, nFET, veriloga

`include "constants.vams"
`include "disciplines.vams"

// Physical Constants
`define pi 3.1416
`define h 4.1357e-015         // planks constant
`define hb 6.5821e-16         // modified plank
                                 constant
`define q 1.602e-19           // Charge
`define epo 8.86e-12          // Permittivity of
                                 free space
`define kb 1.38e-23           // Boltzmann
                                 Constant
`define K 8.612e-5            // Boltzmann
                                 Constant
`define ac 1.43          // C-C Bond Length
`define a 2.48                // lattice constant
                                 in angstroms
`define Vpi 2.97         // C-C Bond energy
`define gamma 0.5
`define Vpp 2.51
`define vf 8.1e15
`define eo 8.85e-22
`define Depth 2
module nFET(b, d, s, g);
inout b;
electrical b;
inout d;
electrical d;
inout s;
electrical s;
input g;
electrical g;
//Instance Parameters
```

```
    parameter real Tox1 = 15e-9;
// oxide layer 1 thickness in meter
    parameter real Tox2 = 120e-9;
// oxide layer 2 thickness in meter
    parameter real T = 300;
// temperature in K
    parameter real L = 250e-9;
// length of CNT in meter
    parameter real Integral=4.1124;
// intergral parameter
    parameter real er1 = 3.9;
// permitivity of oxide layer 1
    parameter real er2 = 3.9;
// permitivity of oxide layer 2
    parameter real Ecrit = 0.2627;
    parameter real Vms = 0;                 // in volts
    parameter real n =11;
    parameter real m = 9;
    parameter real Q01 = 0;
    parameter real Q02 = 0;

    // Variables

    real Vgs,Vds,Vbs,Vgb,Vsb,Vdb; // External voltages
    real Cgs,Cgd,Csb,Cdb,Cgb;     //Capacitance
    real Ids;
    real a1x, a2x,a1y, a2y;       // Lattice Definition
                                     in angstroms
    real L_R;                     // length of R
                                     vector in
                                     angstroms
    real r;                       // Radius in
                                     Angstroms
    real VcbSat;
    real VcbMod;
    real Nc;
    real Ces1;
    real Ces2;
    real Delta;
```

```
    real vfb;
    real Beta;
    real Slope;
    real vth;
    real VcntL;
    real Vcnt0;
    real KT;                    // in eV
    real length;                    // length of CNT in
                                        angstrums
    real Ids_temp;
    real Efs;
    real Efd;
    real Emin;

    analog begin

    Vgs=V(g)-V(s);
    Vds=V(d)-V(s);          //
    Vgb=V(g)-V(b);
    Vsb=V(s)-V(b);
    Vdb=V(d)-V(b);

    Vgs=Vgs+0.5;

    Emin=1;
    length=L*1e10;                  // converted to
                                        angstroms

    KT=$vt;                         // unit
    a2x=a1x=`a*sqrt(3)/2;           // x of a1 and a2 in
                                        angstroms
    a1y=`a/2;                       // y of a1 in
                                        angstroms
    a2y=-`a/2;                      // y of a2 in
                                        angstroms
    L_R=sqrt(pow(n*a1x+m*a2x,2)+pow(n*a1y+m*a2y,2));
//find L_R in angstroms

    r = L_R/`pi/2;                  // Radius in
                                        Angstroms
```

```
    Nc = 4*KT/(`pi*sqrt(3)*`Vpp*`a)*(1e8);
// in 1/cm
    Ces1 =`pi*2*er1*`eo*length/ln((Tox1*1e10+r+sqrt(To
x1*Tox1*1e20+2*Tox1*1e10*r))/r);  // Capacitance in
                                Faradays
    Ces2 =`pi*2*er2*`eo*length/ln((Tox2*1e10+r+sqrt(To
x2*Tox2*1e20+2*Tox2*1e10*r))/r);  // Capacitance in
                                Faradays
    Delta = `q*length*(1e-8)*Nc/Ces1;
              // in volts
    vfb = Vms - Q01/Ces1 - Q02/Ces2 - Q02/Ces1;
              // in volts
    Beta = 0.12*1e3*Ces1/pow(length*(1e-8),2);
    Slope = 0.5*(sqrt(2*Ecrit/KT + 1) - Integral*exp(-
`Depth))/KT;
    vth = vfb + Ecrit - `Depth*KT - Integral*exp(-
`Depth)/Slope;

    //Getting Vcnt0 and VcntL

    Slope = 0.5*(sqrt(2*Ecrit*`Depth/KT + `Depth) -
Integral*exp(-`Depth))/KT;
    VcbSat = Vgs +Vsb - vfb - Ecrit + `Depth*KT +
Integral/Slope*exp(-`Depth);
    //VcbMod = Vgb - Ecrit - vfb - Depth*KT - (Delta/
KT)*sqrt(2*Ecrit*Depth*KT+power((Depth*KT),2));

    if(Vds+Vsb >= VcbSat) begin
       VcntL = Vgs + Vsb - vfb;
    end
    else begin
       VcntL = (Vgs + Vsb - Delta*Integral*exp(-`Depth)
- vfb + Delta*Slope*(Vds+Vsb + Ecrit - `Depth*KT))/(1
+ Delta*Slope);
    end

    Efs=VcntL-Vsb;
    Efd=VcntL-Vds-Vsb;

    Csb=8e-18;
```

```
   Cdb=8e-18;
   Cgb=8e-18;

   if(Vgs >= vth) begin            //saturation region
    Cgs=8e-18;
    Cgd=0;
   end
   else begin                      //linear region
       Cgs=8e-18;
       Cgd=7e-18;
   end

   if(Vgs >= vth) begin            //saturation region
       if(Vsb >= VcbSat)
               Vcnt0 = Vgs + Vsb - vfb;
    else
               Vcnt0 = (Vgs + Vsb -
Delta*Integral*exp(-`Depth) - vfb + Delta*Slope*
(Vsb + Ecrit - `Depth*KT))/(1 + Delta*Slope);

    Ids=((Vgs +Vsb-vfb+KT)*VcntL - 0.5*VcntL*VcntL-
(Vgs +Vsb-vfb+KT)*Vcnt0 + 0.5*Vcnt0*Vcnt0)*Beta;
//Amps
   end
   else begin //linear region
       Vcnt0 = Vgs + Vsb - vfb;
       Ids   =  `q*KT/`pi/`hb*((ln(1+exp(Efs/KT-Emin/
KT))) - (ln(1+exp(Efd/KT-Emin/KT))));
   end

   //current
   I(d,s) <+ Ids*1;
   //Cap
   I(g,s) <+ ddt(Cgs*V(g,s));
   //I(s,g) <+ ddt(Cgs*V(g,s));
   I(g,d) <+ ddt(Cgd*V(g,d));
   //I(d,g) <+ ddt(Cgd*V(g,d));
   I(s,b) <+ ddt(Csb*V(s,b));
   I(d,b) <+ ddt(Cdb*V(s,b));
   //I(g,b) <+ ddt(Cgb*V(g,b));
```

```
end        //analog endmodule


Verilog-AMS code for p-type CNT-FET
// VerilogA for CNTFET, pFET, veriloga

`include "constants.vams"
`include "disciplines.vams"


// Physical Constants
`define pi 3.1416
`define h 4.1357e-015         // planks constant
`define hb 6.5821e-16         // modified plank
                                 constant
`define q 1.602e-19           // Charge
`define epo 8.86e-12          // Permittivity of
                                 free space
`define kb 1.38e-23           // Boltzmann
                                 Constant
`define K 8.612e-5            // Boltzmann
                                 Constant
`define ac 1.43          // C-C Bond Length
`define a 2.48                // lattice constant
                                 in angstroms
`define Vpi 2.97         // C-C Bond energy
`define gamma 0.5
`define Vpp 2.51
`define vf  8.1e15
`define eo 8.85e-22
`define Depth 2

module pFET(b, d, s, g);
inout b;
electrical b;
inout d;
electrical d;
inout s;
electrical s;
input g;
electrical g;
```

```
    //Instance Parameters
    parameter real Tox1 = 15e-9;        // oxide layer
1 thickness in meter
    parameter real Tox2 = 120e-9;       // oxide layer
2 thickness in meter
    parameter real T = 300;             //temperature
in K
    parameter real L = 250e-9;          //  length  of
CNT in meter
    parameter real Integral=4.1124;     //  intergral
parameter
    parameter real er1 = 3.9;           //permitivity
of oxide layer 1
    parameter real er2 = 3.9;           //permitivity
of oxide layer 2
    parameter real Ecrit = 0.2627;
    parameter real Vms = 0;             // in volts
    parameter real n =11;
    parameter real m = 9;
    parameter real Q01 = 0;
    parameter real Q02 = 0;

    // Variables

    real Vsg,Vsd,Vbg,Vbs,Vbd; // External voltages
    real Ids;
    real Cgs,Cgd,Csb,Cdb,Cgb;            //Capacitance
    real a1x, a2x,a1y, a2y;         //Lattice  Definition
in angstroms
    real L_R;               //length  of  R  vector  in
angstroms
    real r;                 //Radius in Angstroms
    real VcbSat;
    real VcbMod;
    real Nc;
    real Ces1;
    real Ces2;
    real Delta;
    real vfb;
    real Beta;
```

```
    real Slope;
    real vth;
    real VcntL;
    real Vcnt0;
    real KT;                    // in eV
    real length;                    // length of CNT in
                                        angstrums
    real Ids_temp;
    real Efs;
    real Efd;
    real Emin;

    analog begin

    Vsg=V(s)-V(g);
    Vsd=V(s)-V(d);          //
    Vbs=V(b)-V(s);
    Vbg=V(b)-V(g);
    Vbd=V(b)-V(d);

    Vsg=Vsg+0.5;
    Emin=1;
    length=L*1e10;                  // converted to
                                        angstroms

    KT=$vt;                         // unit
    a2x=a1x=`a*sqrt(3)/2;           // x of a1 and a2 in
                                        angstroms
    a1y=`a/2;                       // y of a1 in
                                        angstroms
    a2y=-`a/2;                      // y of a2 in
                                        angstroms

    L_R=sqrt(pow(n*a1x+m*a2x,2)+pow(n*a1y+m*a2y,2));
//find L_R in angstroms

    r = L_R/`pi/2;                  // Radius in
                                        Angstroms

    Nc = 4*KT/(`pi*sqrt(3)*`Vpp*`a)*(1e8);
//in 1/cm
```

```
   Ces1 =`pi*2*er1*`eo*length/ln((Tox1*1e10+r+sqrt(To
x1*Tox1*1e20+2*Tox1*1e10*r))/r);
   // Capacitance in Faradays
   Ces2 =`pi*2*er2*`eo*length/ln((Tox2*1e10+r+sqrt(To
x2*Tox2*1e20+2*Tox2*1e10*r))/r);
   // Capacitance in Faradays
   Delta = `q*length*(1e-8)*Nc/Ces1;
            // in volts
   vfb = Vms - Q01/Ces1 - Q02/Ces2 - Q02/Ces1;
            // in volts
   Beta = 0.12*1e3*Ces1/pow(length*(1e-8),2);
   Slope = 0.5*(sqrt(2*Ecrit/KT + 1) - Integral*exp(-
`Depth))/KT;
   vth = vfb + Ecrit - `Depth*KT - Integral*exp(-
`Depth)/Slope;


   //Getting Vcnt0 and VcntL


   Slope  =  0.5*(sqrt(2*Ecrit*`Depth/KT + `Depth) -
Integral*exp(-`Depth))/KT;
   VcbSat = Vsg + Vbs - vfb - Ecrit + `Depth*KT +
Integral/Slope*exp(-`Depth);
   // VcbMod = Vbg - Ecrit - vfb - Depth*KT - (Delta/
KT)*sqrt(2*Ecrit*Depth*KT+power((Depth*KT),2));
   if(Vsd+Vbs >= VcbSat) begin
      VcntL = Vsg + Vbs - vfb;
   end
   else begin
      VcntL = (Vsg +Vbs - Delta*Integral*exp(-`Depth)
- vfb + Delta*Slope*(Vsd+Vbs + Ecrit - `Depth*KT))/(1
+ Delta*Slope);
   end

   Efs=VcntL-Vbs;
   Efd=VcntL-Vsd-Vbs;

   Csb=8e-18;
   Cdb=8e-18;
   Cgb=8e-18;
```

```
if(Vsg >= vth) begin           //saturation region
        Cgs=8e-18;
        Cgd=0;
end
else begin                     //linear region
        Cgs=8e-18;
        Cgd=7e-18;
end

if(Vsg >= vth) begin           //saturation region
        if(Vbs >= VcbSat)
                Vcnt0 = Vsg +Vbs - vfb;
        else
                Vcnt0    =    (Vsg    +Vbs    -
Delta*Integral*exp(-`Depth) - vfb + Delta*Slope*(Vbs +
Ecrit - `Depth*KT))/(1 + Delta*Slope);

        Ids=-((Vsg + Vbs - vfb + KT)*VcntL -
0.5*VcntL*VcntL - (Vsg + Vbs - vfb + KT)*Vcnt0 +
0.5*Vcnt0*Vcnt0)*Beta;       //Amps
end
else begin //linear region
        Vcnt0 = Vsg + Vbs - vfb;
        Ids = -`q*KT/`pi/`hb*((ln(1 + exp(Efs/KT
- Emin/KT))) + (ln(1 + exp(Efd/KT - Emin/KT))));
end

//current
I(d,s) <+ Ids*1;

//Cap
//I(g,s) <+ ddt(Cgs*V(g,s));
I(s,g) <+ ddt(Cgs*V(s,g));
//I(g,d) <+ ddt(Cgd*V(g,d));
I(d,g) <+ ddt(Cgd*V(d,g));
I(b,s) <+ ddt(Csb*V(s,b));
I(b,d) <+ ddt(Cdb*V(s,b));
//I(b,g) <+ ddt(Cgb*V(b,g));

end //analog
endmodule
```

8.3   Implemention of Models

As an example, we have demonstrated here usefulness of Verilog-AMS codes and simulation through Cadence for a CNT-based inverter pair and a five-stage ring oscillator circuit [1]. Figure 8.1a shows the signal response of an inverter pair, which is composed of two series connected inverters. The input signal is 10 GHz square wave with 1 ps rise and fall times. The simulation results show that the delay of the inverter pair is about 3.2 ps. Figure 8.1b shows a plot of the inverter pair average delay versus the supply voltages. At the 0.6 V supply voltage, the average delay is 6.95 ps, which suggests that this inverter pair is able to work for up to 100 GHz input signal.

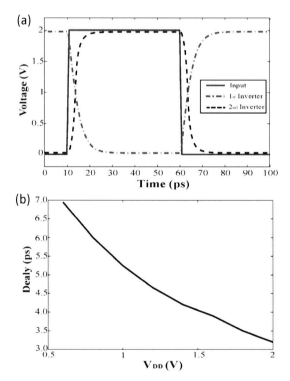

Figure 8.1   (a) Transient behavior of an inverter pair using CNT-FETs (11,9) with V_{fb} = 0 V and ϕ_0 = 0. The dimensions of both the n-type CNT-FET and p-type CNT-FET are as follows: T_{ox1} = 15 nm, T_{ox2} = 120 nm and L = 250 nm and (b) average delay of the inverter pair versus supply voltage.

Figure 8.2a shows the schematic of the five-stage of ring oscillator which was fabricated by Chen et al. [2]. Figure 8.2b shows the simulation result of the ring oscillator output waveform at 0.92 V supply voltage using Verilog-AMS codes. Figure 8.2c shows the oscillation frequency with different supply voltages. The observed derivation of experimental results from the modeled characteristic can be attributed to non-optimized fabrication process, such as the CNT is not perfectly straight. Our model equations and experiments show that the oscillating frequency of this ring oscillator is only about 70–80 MHz at 1.04 V supply voltage. This is due to the CNT-FETs in the ring oscillator, which are 600 nm long, and there are parasitic capacitances associated with the metal wire in the ring oscillator. Shorter length of CNT-FETs will increase the oscillating frequency, as shown in Fig. 8.3. This is due to shorter channel length CNT-FETs, which will give high current and less parasitic capacitances.

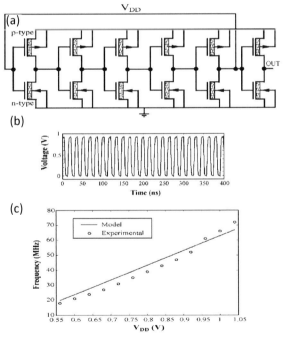

Figure 8.2   (a) Schematic of a five-stage ring oscillator, (b) output waveform, (c) the oscillating frequency versus supply voltage, V_{DD}. The dimensions of both the n-type and p-type CNT-FETs are as follows: $d = 2$ nm and $L = 600$ nm.

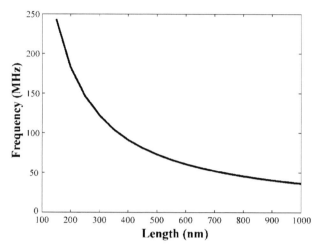

Figure 8.3   Oscillating frequency of the five-stage ring oscillator versus length of the CNT-FETs.

8.4   Summary

In this concluding chapter, we have shown how models developed for the CNT-FETs and CNT interconnections can be easily integrated into commercial EDA tools such as the Cadence through Verilog-AMS codes for analysis and design of emerging carbon nanotube–based integrated circuits.

References

1. Xu, Y., and Srivastava, A. (2009). Dynamic response of carbon nanotube field-effect transistor circuits, *Proceedings of the 2009 NSTI Nanotechnology Conference and Expo*, **1**, pp. 625–628.

2. Chen, Z., Appenzeller J., Solomon P. M., Lin, Y.-M., and Avouris, P. (2006) High performance carbon nanotube ring oscillator, *Proceedings of the 64th Device Research Conference*, pp. 171–172.

Index

VCO, *see* voltage-controlled
  oscillator
very large scale integrated
  (VLSI) 1, 101
VLSI, *see* very large scale
  integrated

VLSI interconnects 58
  next-generation 58
voltage-controlled oscillator
  (VCO) 107, 111, 113–116